Smart Grids: Security and Privacy Issues

Kianoosh G. Boroojeni • M. Hadi Amini
S.S. Iyengar

Smart Grids: Security and Privacy Issues

 Springer

Kianoosh G. Boroojeni
School of Computing and Information
  Sciences
Florida International University
Miami, FL, USA

S.S. Iyengar
School of Computing and Information
  Sciences
Florida International University
Miami, FL, USA

M. Hadi Amini
SYSU-CMU Joint Institute of Engineering
School of Electronics and Information
  Technology
Sun Yat-sen University
Guangzhou, Guangdong, China

Department of Electrical and Computer
  Engineering
Carnegie Mellon University
Pittsburgh, PA, USA

ISBN 978-3-319-83197-8   ISBN 978-3-319-45050-6   (eBook)
DOI 10.1007/978-3-319-45050-6

Printed on acid-free paper

This Springer imprint is published by Springer Nature
The registered company is Springer International Publishing AG
The registered company address is: Gewerbestrasse 11, 6330 Cham, Switzerland

Preface

There has been a growing trend in the power systems from a centralized producer-driven grid to a smarter interactive customer network. This requirement compels a new way of designing smart grids for a more reliable and secure power system performance. Involving the demand side in the power system management requires large-scale utilization of distributed communication networks. In this context, there is a considerably increasing concern regarding the security and privacy of both physical and communication layers of the network. This book utilizes an advanced interdisciplinary approach to address the existing security and privacy issues and propose legitimate countermeasures for each of them in the standpoint of both computing and electrical engineering. The proposed methods are theoretically proved by mathematical tools and illustrated by real-world examples.

This book paves the way for researchers working on privacy and security issues spread throughout computer science and smart grids. Furthermore, it provides the readers with a comprehensive insight to understand an in-depth big picture of privacy and security challenges in both physical and information aspects of smart grids. This book can be used as textbook for graduate-level courses in computer engineering, computer science, electrical engineering, and other related areas.

Features

Here are the unique aspects of our book which address the oblivious network routing problems:

(1) Preserving of the location privacy of mobile users of future smart grids.
(2) The security of smart grids attracts researchers' attentions.
(3) Information privacy for smart meters and mobile users plays a key role in future smart grids.
(4) Maintaining the reliability of the smart grids.

(5) Providing security countermeasures for false data injection into the communication networks of smart grids.
(6) Evaluating the privacy and security of both physical and information infrastructures.

Proposing a novel congestion-based economic dispatch framework based on oblivious routing network concept.

Intended Audience

This monograph is suitable for senior undergraduate students, graduate students, and the researchers working in the related areas.

Miami, FL, USA   Kianoosh G. Boroojeni
Pittsburgh, PA, USA   M. Hadi Amini
Miami, FL, USA   S.S. Iyengar

Acknowledgments

This work has evolved from our research on security and privacy in smart grids. Kianoosh G. Boroojeni and S.S. Iyengar would like to thank NSF, NIH, and other research funding institutes for funding their research works.

M. Hadi Amini would like to thank his parents, Alireza and Parvin, for their unconditional support and encouragement throughout his life. He also would like to appreciate the kindly support of his artist sister, his uncle, and his brother-in-law. He also would like to express his gratitude towards his beloved niece.

Kianoosh G. Boroojeni would like to express his sincere gratitude to his family for their continuous inspiration and support.

Contents

Biography

Kianoosh G. Boroojeni

Kianoosh Gholami Boroojeni is a Ph.D. candidate of computer science at FIU. He received his B.Sc. in computer science in the University of Tehran, Iran (2012). His research interests include smart grids and algorithms.

During Kianoosh's graduate years, he has authored a book at MIT Press, a book at Springer Publishers, and several journal and conference papers. Currently, Kianoosh is collaborating with Dr. S.S. Iyengar on some cybersecurity issues in the context of smart grids and cloud computing.

M. Hadi Amini

Mohammad Hadi Amini received the B.Sc. degree from Sharif University of Technology, Tehran, Iran, in 2011, and the M.Sc. degree from Tarbiat Modares University, Tehran, in 2013, both in electrical engineering. He also received the M.Sc. degree in electrical and computer engineering from Carnegie Mellon University in 2015. He is currently pursuing the dual-degree Ph.D. in electrical and computer engineering with the Department of Electrical and Computer Engineering, Carnegie Mellon University (CMU), Pittsburgh, PA, USA, and Sun Yat-sen University-CMU Joint Institute of Engineering, Guangzhou, China. He is also with the School of Electronics and Information Technology, SYSU, Guangzhou, China, and SYSU-CMU Shunde International Joint Research Institute, Shunde, Guangdong, China. Hadi serves as reviewer for several high-impact journals and international conferences and symposiums in the field of smart grid. He has published more than 30 refereed journals and conference papers in the smart grid-related areas and served as a session chair in INFORMS Annual Meeting 2015. He has been awarded the 5-year scholarship from the SYSU-CMU Joint Institute of Engineering in 2014, sustainable mobility summer fellowship from Massachusetts Institute of Technology (MIT) Office of Sustainability in 2015, and the dean's honorary award from the president of Sharif University of Technology in 2007. His current research interests include smart grid, electric vehicles, optimization methods in interdependent power and transportation networks, and distributed optimization.

S.S. Iyengar

S.S. Iyengar is a leading researcher in the fields of distributed sensor networks, computational robotics, and oceanographic applications and is perhaps best known for introducing novel data structures and algorithmic techniques for large-scale computations in sensor technologies and image processing applications. He has published more than 500 research papers and has authored or coauthored 12 textbooks and edited 10 others. Iyengar is a member of the European Academy of Sciences, a fellow of the Institute of Electrical and Electronics Engineers (IEEE), a fellow of National Academy of Inventors (NAI), a fellow of the Association for Computing Machinery (ACM), a fellow of the American Association for the Advancement of Science (AAAS), and fellow of the Society for Design and Process Science (SDPS). He has received the Distinguished Alumnus Award of the Indian Institute of Science. In 1998, he was awarded the IEEE Computer Society's Technical Achievement Award and is an IEEE Golden Core Member. Professor Iyengar is an IEEE Distinguished Visitor, SIAM Distinguished Lecturer, and ACM National Lecturer. In 2006, his paper entitled "A Fast Parallel Thinning Algorithm for the Binary Image Skeletonization", was the most frequently read article in the month of January in the *International Journal of High Performance Computing Applications*. His innovative work called the Brooks-Iyengar algorithm along with Prof. Richard Brooks from Clemson University is applied in industries and some real-world applications.

Chapter 1
Overview of the Security and Privacy Issues in Smart Grids

In recent years, there is an increasing trend in the power systems from a centralized fossil fuel-based grid toward a distributed green-based network. This requirement compels a new way of designing smart grids for a more reliable and secure power system performance. Involving the demand side in the power system management requires large-scale utilization of distributed communication networks. In this context, there is a considerably increasing concern regarding the security and privacy of both physical and communication layers of the network. This book utilizes an advanced interdisciplinary approach to address the existing security and privacy issues and propose legitimate countermeasures for each of them in the standpoint of both computer science and electrical engineering. The proposed methods are theoretically proved by mathematical tools and illustrated by real-world examples.

By adding new functionality and communication capabilities to the power system, the smart grid infrastructure can be more efficient, more resilient, and more affordable to manage and operate. The communication infrastructure in smart grid has the potential of introducing new security vulnerabilities into the system. To achieve security requirements in the smart grid, we require comprehensive security mechanisms that (a) preserve the privacy of the customers and utility companies; (b) prevent malicious attacks targeting availability; (c) prevent undetected modification of information by unauthorized persons or systems; (d) distinguish between legitimate and illegitimate users based on authentication; (e) prevent access to the system by other systems without permission.

1.1 Security Issues in Smart Grid

Recent advances in wireless communications and smart grid designs have enabled the development of renewable-based power distribution systems [1–3]. In a related context, the negative effects of coal plants will create a major environmental hazard

© Springer International Publishing Switzerland 2017
K.G. Boroojeni et al., *Smart Grids: Security and Privacy Issues*,
DOI 10.1007/978-3-319-45050-6_1

in addition to changing the design of smart distribution networks [4]. Henceforth, not only the distribution of electric power approach is changing but also the producers of the electricity are focusing on large-scale utilization of distributed renewable resources (DRRs). Distributed generations can also be utilized to overcome some issues related to the smart distribution networks, such as congestion management and reactive power control [5].

Smart power grids deploy several technologies and methods to obtain a more reliable and secure grid, including demand response programs [6, 7], electric vehicles [8], distributed optimization [9], renewable resources [10], and smart home appliances' scheduling [11]. One of the most critical challenges of designing smarter grids is to make the power distribution facilities more flexible to the variable nature of the electricity consuming and stochastic generation pattern of DRRs [12, 13]. Operational challenges of high penetration of the renewable resources into the power grid are addressed in [14]. More specifically, producing power using a green plant such as wind farms doesn't guarantee a low-threshold of energy throughput in a specific time frame. In addition, the consumers have a time-varying usage pattern in the context of an electricity residential system [15]. In order to solve the problem of increasing intermittency of DRRs and demand, a comprehensive stochastic reactive power management scheme is introduced in [16]. According to [17], a decentralized control approach is proposed to enable demand side management by following the desired set-points for distributed level controllers. Moreover, a distributed control algorithm is introduced in [18] to optimize the generation cost in smart micro grids with DRRs. By considering all of these uncertainties, designing a versatile power distributing scheme is absolutely fundamental for developing a reliable grid. In other words, the goal is to create a reliable power distribution grid which can satisfy the consumers by providing electricity adequately for them in a real-time manner. According to [19], an approximate model of distribution network is utilized to facilitate the problem of loss minimization into the linearly constrained convex optimization problem. Furthermore, distributed energy management plays a pivotal role in future power systems. In [20], a comprehensive review of distributed energy management methods in smart grid is performed.

The problem of allocating energy from renewable sources to flexible consumers in electricity markets was investigated by Neely et al. in 2010 [21]. The power generated by the green resources in their work is modeled as a time-varying supply process and the energy distribution cost function in their problem has been assumed to be linear or dynamic according to the suggested energy price. In [22], a novel approach for the distribution of green power resources in future power systems is introduced. In [23], we proposed a two-tier forecaster for stochastic generation resources which facilitates solving the economic dispatch of intermittent DRRs in future distribution networks. Furthermore, distributed sensor networks are studied from a mathematical perspective in [24]. These networks can facilitate the monitoring process in the smart grid management and control.

1.2 Physical Network Security

In recent years, there is an ever-increasing concern about energy consumption and its environment issues, reliable energy supply, and sustainable development of energy and power networks. These issues motivate the evolution of Smart Grid (SG) as a novel means to worldwide electricity grid [1]. Smart grid includes several elements, such as electric vehicles [25–27], distributed renewable resources [28], smart appliances [29], distributed state estimation and smart load management [2, 20], and demand response [6]. In [26] a comprehensive method is proposed for the allocation of electric vehicle parking lots in the smart power distribution network.

Here, we provide a brief overview of related studies on reliability of power systems. Electricity grid infrastructures have been developed long ago and there are huge populations of components (e.g., generating units, transformers, circuit breakers, overhead lines, etc.) across the system with the ages of over 25–40 years in service playing critical roles in power system operation. Utilities, thus, have to anticipate investment waves to appear as a critical concern in their current asset management practices. Assessing the reliability of the equipment and their failure impact on the electricity grid performance over time is a challenge, but a necessity. Accordingly, maintenance planning and scheduling of the equipment needs to be in line with their individual and system reliability indicators. Research on reliability analysis of the system and implementation of advanced maintenance strategies has been conducted in all the hierarchical levels of electric power systems: on generating units and power plants [30], on transmission grid equipment [31–33], and on distribution systems [34, 35]. It has been proved that those equipments that reveal a higher risk to the safe and reliable operation of the grid are the ones needing maintenance attentions the most and should be on top of the priority list for maintenance and resource allocation decisions.

In the future SG, it is critical to take advantage of the advancements in communication technologies to enable automated and intelligent system management [36]. Although the currently available communication technologies can greatly satisfy our personal communication needs, applying them to power systems and addressing the specific requirements for power communications are challenging. In particular, the challenges can be in different applications in distribution networks such as load management and demand response [37–39], electric vehicle charging and discharging management [40], accurate demand forecasting tools [41], as well as a variety of energy management problems in transmission networks such as optimal power flow [42], and power dispatch with high penetration of renewable energies [43, 44]. Numerous potential vulnerabilities to cyber-attacks arise [45]. The cyber security requirements and the possible vulnerabilities in smart grid communications and the current security solutions can be found in [45]. In [46a], system-of-systems concept has been deployed to deal with power distribution network planning. The increasing penetration of microgrids in the electricity markets can pose challenges to the system operation. Distribution markets are new concepts put forward to facilitate efficient and market-based integration of microgrids in the future smart grids [46b]. Furthermore, a comprehensive planning model for power distribution networks was introduced in [46c]. The security in smart grid can be categorized into five categories [47, 48]: process control system security, smart meter security, power system state estimation security, smart grid communication protocol security, and smart grid simulation for security analysis.

SG will gradually upgrade the power systems to achieve a more reliable, environmentally-friendly, and economically-operating system. A probabilistic model for electric vehicle parking lot modeling is proposed in [49]. This method is based on the probabilistic arrival and departure time and the expected driven distance of single electric vehicles. Self-healing and autonomous decision making methods play a pivotal role in the future power systems as complex dynamical networks [4, 50, 51]. A comprehensive study of the smart grid effects on the smart distribution network design is provided in [4].

Bulk electric transmission systems have been traditionally characterized with static assets and fixed configuration over time except in the cases of faults and forced outages. Power system topology control, often called transmission switching, offers the system operators an opportunity to harness the flexibility of the transmission system topology by temporarily removing lines out of service. With minimum additional cost and by changing the way how electricity flows through the system, transmission switching can be employed either in emergency scenarios (to alleviate violations, congestions, overloading conditions, and even load shed recovery) [52, 53] or during normal operating conditions (for higher economic benefits). Several studies have been conducted on transmission switching and advanced decision making frameworks have been developed for practical implementation of this technology more frequently [54–56]. Such considerations, which are employed in the operational time frame, make it possible to have more efficient use of the existent network facilities.Though being performed for decades on a very limited scale with rather focused aims, transmission switching has gained further importance with the increased penetration of renewable energy resources and the growing demand for more reliable operation of power systems [56].

In [51], multi-agent systems are utilized to develop a smart load management strategy for the smart distribution networks. According to [51], multi-agent systems include multiple autonomous agents (including software and hardware) with diverging information or interest [57]. These systems are widely used in different applications, from small scale systems which are used for personal purposes, to complex large-scale applications for industrial purposes [58]. According to [51, 59], an agent has the following features:

- Autonomy: ability to control its actions and internal state without the direct intervention of human or other agents;
- Social-ability: agents can interact with human and other agents utilizing agent communication language;
- Reactivity: agents can analyze the environment and respond to the corresponding changes that happens in it;
- Pro-activeness: the response of agents to the environment is in object-oriented direction.

Optimal operation of microgrids and the smart grids is obtained by developing an energy management system in [60]. The proposed method is based on hybrid connected neural networks and optimal power flow.

Power flow problem is solved to obtain the steady-state operation point of the power systems. In the optimal power flow problem, contingencies (fault, congestion, physical attack, generator failure) and physical limits of network are considered as the constraints while solving the power flow problem. Furthermore, the contingency analysis plays an important role in the smart grid. In [61], an interruption prediction approach is proposed to improve the reliability of the smart grid using the historical weather data and the chronological interruption data. To this end, Sarwat et al. used artificial neural networks to perform an accurate forecast based on the available historical data and the effectiveness of their proposed method is validated via real-world historical data from Florida Power and Light (FPL) as well as National Climatic Data Center (NCDC). Furthermore, the real-time contingency data can be used to prevent the issues caused by outages. Smart grids require a robust and secure self-decision making ability. To this end, power flow calculations should be performed accurately. Any computation errors due to unintentional faults or bad data injection to the system in the power flow calculation may lead to a major outage or black-out [62].

The ultimate objective of solving power flow problem is to determine the amount of voltage phase and voltage angles, active power, and reactive power in the power system. Four widely used methods in the power flow analysis can be listed as [63, 64]:

- **Gauss–Seidal method** is based on an iterative method for solving linear systems. According to the literature, this approach has been widely applied in the problems where the system representation matrix is nonsingular, i.e., the determinant of matrix is non-zero. The matrix should be diagonally dominant, or symmetric and positive definite so that the convergence of method is guaranteed.
- **Newton Raphson method** is another technique for solving power flow problem. The convergence of this method is quadratic. However this method is used in the literature for power flow studies, there are some issues with the method.
- **Fast Decoupled method**: Fast power flow algorithms are used to expedite the power flow computation into short durations (seconds or a proportion of seconds).
- **DC power flow method**[65]: Further simplifications such as neglecting the $Q-V$ equation and assuming the voltage magnitudes are constant at 1.0 per-unit will lead to DC power flow formulation [66]. This simplification and the equation related to the DC power flow will be elaborately discussed in a separate section. In [67], some of the advantages of DC power flow are mentioned as follows:

  (*i*) Non-iterative, reliable, and unique solutions;
  (*ii*) Simplified methods and software;
  (*iii*) Efficient solution and optimization, specifically for contingency analysis;
  (*iv*) Minimal and easy-to-obtain network data requirement;

(*v*) Linear structure that matches the economic theory for transmission-oriented market studies;

(*vi*) Fairly accurate approximation of active power flows.

Some studies tried to analyze the theoretical error of DC models in power flow [68, 69]. Hybrid AC/DC microgrids control requires DC power-flow method for studying the technical aspects [70].

1.3   Information Network Security

Using cloud computing applications for big data analytics in smart grid plays an important role in dealing with the heterogeneous structure of the smart grids [71, 72]. According to [73], cloud computing model that meets the requirements of data and computing intensive applications of the smart grid can be widely used in future power system computations. Additionally, big data analytics have been widely deployed to improve both power systems operation and protection [74]. In [74], three aspects of using large data sets in power systems are studied comprehensively: feature extraction, systematic integration for power system applications, and examples of typical applications in the utility industry. Moreover, cloud computing has become a growing architectural model for organizations seeking to decrease their computer systems maintenance costs by migrating their computational tasks to third party organizations who offer software-as-a-service, platform-as-a-service, etc. As the result of such transition, many critical security concerns have emerged among the clients of clouds including data blocking and data leakage in the form of Distributed Denial of Service (DDoS) attacks [75]. As the cloud customers may have possibly thousands of shared resources in a cloud environment, the likelihood of DDoS attack is much more than a private architecture [76].

One of the main goals of DDoS attacks in cloud environments is to exhaust computer resources especially network bandwidth and CPU time so that the cloud service gets unavailable for the real legitimate clients [76]. In a general DDoS attack, the attacker usually mimics the legitimate web data traffic pattern to make it difficult for the victims to identify the attackers. This type of attack called "zombie attack" is a common type in cloud computing and is widely considered to be successful in circumventing the big portion of attack detection algorithms which work based on the abnormality of the traffic pattern generated by the DDoS attackers [76–78].

In recent years, many attempts have been made to mitigate the DDoS zombie attacks in general or specifically in the cloud environments. As we will see later in the literature review, these attempts have all relied on the conventional Internet routing algorithms and protocols (e.g., link-state, distance-vector) to route the data flow through the Internet. To the best of our knowledge, this is the first try ever made to mitigate the DDoS attacks by not utilizing the conventional Internet routing algorithms and creating a novel overlay network routing scheme based on the oblivious network design principles and algorithms. This project mainly focuses on the mitigation of the zombie attack and its destructive effect on the cloud services. The proposed mechanism is compatible with many of the DDoS detection and mitigation algorithms and can be integrated with them in order to enhance the security.

Related Work   Here, we review some of the recent studies on how to detect and mitigate the DDoS attacks in general and specifically in cloud environments.

1.3.1   Detection Mechanisms

The most common type of DDoS attacks in Internet is flooding. This popular attack has made many attempts in order to create effective countermeasures against it [79–81]. The designed/implemented countermeasures have an attack detection algorithm which predicts the occurrence of attacks utilizing the fact that the pattern of data traffic made by the hackers is deviated from the normal (natural) traffic pattern. The greater this deviation is, the higher the precision of the attack detection algorithms are. In 2015, Wang et al. proposed a graphical model in order to detect the security attacks (distributed denial of service) and inferring the probability distribution of attack [82].

As a response to this type of counter attacks, the attackers may regulate their generated data traffic pattern in a way that it no longer distinguishable with the other network data traffic and confuses the attack detection algorithms. Zombie attack is a good example in which the attackers perform DDoS attacks through a Poisson process in order to confuse the detection algorithms [77, 78]. Joshi et al. designed a cloud trace back model in dealing with DDoS attacks and addressed its performance using back propagation neural network and experimentally showed that the model is useful in tackling DDoS attacks [76].

1.3.2   Mitigation Mechanisms

In 2013, Mishra et al. proposed Multi-tenancy and Virtualization as two major solutions to mitigate the DDoS attacks in cloud environments [83]. Zissis et al. proposed a security solution to the security issues which helps the cloud clients make sure of their security by trusting a Third Party in a way that trust is created in a top-down manner where each layer of the cloud system trusts the layer lying immediately below it. In 2011, Lua et al. reduced the possibility of DDoS attacks by utilizing a transparent and intelligent fast-flux swarm network which is a novel, efficient, and secure domain naming system.

This chapter proposes an Oblivious Routing-based Security Scheme ($\mathcal{ORSS}$) which mitigates the DDoS attack and related performance issues effectively while routing data through the Internet connecting the cloud servers and customers together. An oblivious routing algorithm usually proposes an overlay network in the form of a spanning tree or a set of trees on the graph representing the network [22]. These schemes are usually pretty versatile to the time-varying network topology and dynamic traffic pattern between different source–sink pairs by routing the traffic flows in a distributed way over the network and preventing the edge/node congestion. Moreover, the routing cost in such algorithms is usually proved to be asymptotically bounded to some coefficient of the optimal/minimum cost; this coefficient which is called the *competitiveness ratio* of the algorithm (or its corresponding routing scheme) has a value of greater than one [22, 84].

By defining the routing cost of data flow as a representation of security attack risk, we utilize an oblivious routing algorithm of competitiveness ratio $O(\log^2 |V|)$ mentioned in [22] to design a secure overlay network routing scheme. The proposed security scheme is mathematically proved to guarantee some security low-threshold in long-term run. As the security attacks cause reduction in performance, we will show how $\mathcal{ORSS}$ enhances the system performance. The simulation results will support the theoretical bound.

1.4 Privacy Issues in Smart Grids

Continued advances in positioning technologies and mobile devices like wireless sensors and smart phones have increased the demand for location-based services (LBSs). One major concern in the wide deployment of LBSs is how to preserve the location privacy of users while providing them with a service based on their locations. LBS providers can be victims of the privacy attackers to track some users, or they themselves may abuse the users' location information for advertising purposes [85–87].

Consider a mobile image sensor flying over an area to keep watch over an area (e.g., patrol borders or battlefields). To do this, it moves to different places and frequently uses some LBS to gather (visual) information about its current location, etc. In many cases, the sensor as a patrol node prefers to randomly walk over the patrolled area as moving regularly degrades the Quality of Service (QoS) of patrolling. In fact, random walk makes the adversary unable to predict the mobility pattern of the sensor; i.e. the sensor nodes may utilize random walk in order to protect their location privacy. However, as the randomly walking sensors may need to expose their location to some LBS, their location privacy may be threatened by the LBS itself or some third party (adversary). Additionally, Shi et al. [88] showed that random walk is not truly memory-less. In fact, merely using a random walk doesn't guarantee the preservation of location privacy. These issues urge us to design a location privacy-preserving mechanism for randomly walking devices to hide their instantaneous location and movement trajectory.

Related Work   There are two general approaches dealing with the location privacy issue in LBSs. In this section, firstly we address the couple of approaches; then, we explain how to quantify the location privacy of LBS users. Finally, we will focus on our contribution.

1.4.1   k-Anonymity Cloaking

In this approach, instead of sending one single user's LBS request to the server, including her exact location, k-anonymity cloaking employs a trusted third party who collects k neighboring users' requests and sends them all together to the LBS service provider. This approach doesn't address the case that the user density is high; in this case, the k users' locations may be very close to each other, and hence this approach will still reveal the user's location privacy to some extent. This approach was originally proposed by Gruteser and Grunwald [89]. Their work may lead to large service delay if there are not enough users requesting LBSs. Later on, Gedik and Liu [90] designed a joint spatial and temporal cloaking algorithm which collects k LBS requests, each from a different user in a specified cloaking area within a specified time period and then sends them to the service provider. A negative point in their work is that if there are only less than k requests within the predefined time period, the users' requests will be blocked.

In 2009, Meyerowitz and Choudhury [91] tried to improve the service accuracy by predicting the users' paths and LBS queries, and send the results to users' before they submit queries. The main drawback of their approach is the network delay occurred because of high communication overhead. For more treatment on k-anonymity cloaking, see [92–97].

1.4.2   Location Obfuscation

We can divide the solutions with this approach into two categories: solutions that preserve the location privacy of the users by inserting some fake LBS requests; and those which deviate a user's location from the real one in her LBS request to protect her location privacy.

As examples of the solutions in the earlier category, consider the schemes proposed by Kido et al. [98], Lu et al. [99], and Duckham and Kulik [100]. In these schemes, the user generates some fake locations (dummies) using some dummy generation methods and submits the dummies and its own location to the LBS server. The server analyzes every submitted query and replies properly. The major drawback of the solutions in this category is that the server is used inefficiently and may become the system bottle-neck. Additionally, users' location privacy is not preserved in advance.

The second category of solutions like the ones proposed by Ardagna et al. [101], Pingley et al. [102] and Damiani et al. [103] hide users' real locations, e.g., by submitting shifted locations. Such schemes trade service accuracy for location privacy.

Finally, in 2013, Ming Li et al. [86] proposed a location privacy-preserving scheme called *n*-CD which doesn't use a third party and provides a trade-off between the privacy level and the system accuracy (concealing cost).

1.4.3 Location Privacy Quantification and Formalization

In 2011, Shokri et al. [104, 105] formalized localization attacks using Bayesian inference for Hidden Markov Processes. They mathematically modeled a location privacy-preserving mechanism, users mobility pattern and adversary knowledge-base. They also divide the LBSs into two classes: those who sporadically ask their users to expose their location, and those who continuously do. Additionally, they quantified the user location privacy as the expected distortion of adversary's guess from the reality of user's location. They used this quantification method to evaluate the effectiveness of location obfuscation and fake location injection mechanisms in the improvement of location privacy level.

1.5 Book Structure and Outlook

This book discusses the security and privacy issues of smart grids and their possible countermeasures in three separate parts. Part I focuses on physical network security challenges facing the wide implementation of smart grids. This part consists of Chaps. 2, 3, and 4. Chapter 2 addresses the reliability issues in smart grids. In Chaps. 3 and 4, data falsification, which is a significant security attack in smart grids, is studied thoroughly with the emphasis on power flow estimation and bad data detection as two of its effective countermeasures.

Part II focuses on the information network security of smart grids in the context of cloud environment. Due to significant advances of sensor networks and computer devices in recent years, smart grids use big data analysis in order to operate more effectively. However, because of economical and reliability reasons, dealing with big data storage and analysis with the classic client–server architecture of computing services is not feasible anymore. In such conditions, there is an increasing transition toward client environments which offers a more customer-driven, inexpensive, and reliable services in comparison with client–server computer architectures. Although cloud environments have significant reliability and economical benefits, wide use of them will make the computer systems prone to many new security challenges including the threat of data interception, sniffing and manipulation, data privacy, etc. Part II offers some effective countermeasures for fighting against such attacks.

Part III which consists of Chaps. 6 and 7 addresses the privacy issues facing the wide deployment of smart grids in today's World. Chapter 6 focuses on end-user data privacy issues and offers effective ways to deal with them in real World. Moreover, Chap. 7 discusses another privacy issues for mobile users and the possibility of location and trajectory privacy threats evolving when new technologies like E2X get implemented and widely used by vehicles.

References

1. V.C. Gungor et al., Smart grid technologies: communication technologies and standards. IEEE Trans. Ind. Inf. **7**(4), 529–539 (2011)
2. National Institute of Standards and Technology, NIST framework and roadmap for smart grid interoperability standards, release 1.0, Office of the National Coordinator for Smart Grid Interoperability-U.S. Department of Commerce, NIST Special Publication 1108, Jan 2010
3. S.M. Kay, *Fundamentals of Statistical Signal Processing: Estimation Theory*, 1st edn. (Prentice-Hall International Editions, Englewood Cliffs, 1993)
4. R.E. Brown, Impact of smart grid on distribution system design, in *Proceedings of IEEE Power and Energy Society General Meeting*, Pittsburgh, PA, July 2008, pp. 1–4
5. A. Elmitwally, M. Elsaid, M. Elgamal, Z. Chen, A fuzzy-multiagent self-healing scheme for a distribution system with distributed generations. IEEE Trans. Power Syst. **99**, 1–11 (2014)
6. F. Kamyab, M.H. Amini, S. Sheykhha, M. Hasanpour, M.M. Jalali, Demand response program in smart grid using supply function bidding mechanism, IEEE Transactions on Smart Grid **7**(2), 1277–1284 (2016)
7. S. Mhanna, A.C. Chapman, G. Verbič, A fast distributed algorithm for large-scale demand response aggregation. IEEE Trans. Smart Grid **7**(4), 2094–2107 (2016)
8. M.H. Amini et al., ARIMA-based demand forecasting method considering probabilistic model of electric vehicles' parking lots, in *Proceedings of IEEE Power and Energy Society General Meeting*, 2015, pp. 1–5
9. M.H. Amini, R. Jaddivada, S. Mishra, O. Karabasoglu, Distributed security constrained economic dispatch, in *IEEE PES Innovative Smart Grid Technologies Conference (ISGT-Asia 2015)*, Bangkok, 3–6 Nov 2015
10. A. Mohsenzadeh, M.-R. Haghifam, Simultananeus placement of conventional and renewable distributed generation using multi objective optimization, in *Proceedings of IEEE, Integration of Renewables into Distributed Grid Workshop, CIRED*, 2012
11. M.H. Amini, J. Frye, M.D. Ilic, O. Karabasoglu, Smart residential energy scheduling utilizing two stage mixed integer linear programming, in *IEEE 47th North American Power Symposium (NAPS 2015)*, Charlotte, NC, 4–6 Oct 2015
12. A. Papavasiliou, S.S. Oren, Supplying renewable energy to deferrable loads: algorithms and economic analysis, in *Proceedings of IEEE Power and Energy Society General Meeting*, Minneapolis, Minnesota, July 2010, pp. 1–8
13. A.R. Di Fazio, G. Fusco, M. Russo, Enhancing distribution networks to evolve toward smart grids: The voltage control problem, in *Proceedings of IEEE 52nd Annual Conference on Decision and Control (CDC)*, Firenze, Dec 2013, pp. 6940–6945
14. J.M. Morales et al., *Integrating Renewables in Electricity Markets: Operational Problems*. vol. 205 (Springer Science & Business Media, Berlin, 2013)
15. W. Tushar et al., Three-party energy management with distributed energy resources in smart grid. IEEE Trans. Ind. Electron. **62**(4), 2487–2498 (2015)
16. V. Kekatos, G. Wang, A.J. Conejo, G.B. Giannakis, Stochastic reactive power management in microgrids with renewables. IEEE Trans. Power Syst. **99**, 1–10 (2014)

17. S.E. Shafiei et al., A decentralized control method for direct smart grid control of refrigeration systems, in *Proceedings of IEEE 52nd Annual Conference on Decision and Control (CDC)*, Firenze, Dec 2013, pp. 6934–6939
18. G. Cavraro, R. Carli, S. Zampieri, A distributed control algorithm for the minimization of the power generation cost in smart micro-grid, in *Proceedings of IEEE 53rd Annual Conference on Decision and Control (CDC)*, Los Angeles, CA, Dec 2014, pp. 5642–5647
19. S. Bolognani, S. Zampieri, A distributed control strategy for reactive power compensation in smart microgrids. IEEE Trans. Autom. Control **58**(11), 2818–2833 (2013)
20. S. Kar, G. Hug, J. Mohammadi, J.M.F. Moura, Distributed state estimation and energy management in smart grids: a consensus+innovations approach. IEEE J. Sel. Top. Sign. Proces. **99**, 1–16 (2014)
21. M.J. Neely, A. Tehrani, A. Dimakis, Efficient algorithms for renewable energy allocation to delay tolerant consumers , in *Proceedings of IEEE First International Conference on Smart Grid Comm. (SmartGridComm)*, Gaithersburg, MD, Oct 2010, pp. 549–554
22. S.S. Iyengar, K.G. Boroojeni, *Oblivious Network Routing Algorithms and Applications* (MIT Press, Cambridge, 2015)
23. K.G. Boroojeni, S. Mokhtari, M.H. Amini, S.S. Iyengar, Optimal two-tier forecasting power generation model in smart grid. Int. J. Inf. Process. **8**(4), 1–10 (2014)
24. S.S. Iyengar, K.G. Boroojeni, N. Balakrishnan, *Mathematical Theories of Distributed Sensor Networks* (Springer Publishing Company, Incorporated, Berlin, 2014)
25. W. Su, H.R. Eichi, W. Zeng, M.-Y. Chow, A survey on the electrification of transportation in a smart grid environment. IEEE Trans. Ind. Inf. **8**(1), 1–10 (2012)
26. M.H. Amini, A. Islam, Allocation of electric vehicles' parking lots in distribution network, in *Proceedings of IEEE Innovative Smart Grid Technologies Conference (ISGT)*, Washington, DC, Feb 2014, pp. 1–5
27. M.H. Amini, A.I. Sarwat, Optimal reliability-based placement of plug-in electric vehicles in smart distribution network. Int. J. Eng. Sci. **4**(2), 43–49 (2014)
28. J.M. Carrasco et al., Power-electronic systems for the grid integration of renewable energy sources: a survey. IEEE Trans. Ind. Electron. **53**(4), 1002–1016 (2006)
29. X. Yu, C. Cecati, T. Dilon, M.G. Simoes, The new frontier of smart grids. IEEE Ind. Electron. Mag. **5**(3), 49–63 (2011)
30. T. Konstantin, et al. Local control of reactive power by distributed photovoltaic generators, in *First IEEE International Conference on Smart Grid Communications (SmartGridComm)* (2010)
31. R. Ghorani, M. Fotuhi-Firuzabad, P. Dehghanian, W. Li, Identifying critical component for reliability centered maintenance management of deregulated power systems. IET Gener. Transm. Distrib **9**(9), 828–837 (2015)
32. P. Dehghanian, T. Popovic, M. Kezunovic, Circuit breaker operational health assessment via condition monitoring data, in *46th North American Power Symposium, (NAPS)*, Sep 2014 (Washington State University, Pullman, Washington, 2014)
33. P. Dehghanian, M. Moeini-Aghtaie, M. Fotuhi-Firuzabad, R. Billinton, A practical application of the Delphi method in maintenance-targeted resource allocation of distribution utilities, in *13th International Conference on Probabilistic Methods Applied to Power Systems, PMAPS*, July 2014 (Durham University, Durham, 2014)
34. P. Dehghanian, M. Fotuhi-Firuzabad, F. Aminifar, R. Billinton, A comprehensive scheme for reliability centered maintenance implementation in power distribution systems- Part I: methodology. IEEE Trans. Power Delivery **28**(2), 761–770 (2013)
35. P. Dehghanian, M. Fotuhi-Firuzabad, S. Bagheri-Shouraki, A.A. Razi Kazemi, Critical component identification in reliability centered asset management of distribution power systems via fuzzy AHP. IEEE Syst. J. **6**(4), 593–602 (2012)
36. S. Bahrami, M. Parniani, A. Vafaeimehr, A modified approach for residential load scheduling using smart meters, in *IEEE PES Innovative Smart Grid Technologies Europe (ISGT Europe)*, Berlin, 2012

37. S. Bahrami, V.W.S. Wong, An autonomous demand response program in smart grid with foresighted users, in *Proceedings of IEEE SmartGridComm*, Miami, FL, 2015
38. S. Bahrami, F. Khazaeli, M. Parniani, Industrial load scheduling in smart power grids, in *22nd International Conference and Exhibition on Electricity Distribution (CIRED 2013)*, Stockholm 2013, pp. 1–4
39. P. Samadi, H. Mohsenian-Rad, V.W.S. Wong, R. Schober, Real-time pricing for demand response based on stochastic approximation. IEEE Trans. Smart Grid **5**(2), 789–798 (2014)
40. S. Bahrami, M. Parniani, Game theoretic based charging strategy for plug-in hybrid electric vehicles. IEEE Trans. Smart Grid **5**(5), 2368–2375 (2014)
41. K.G. Boroojeni, M.H. Amini, S. Bahrami, S.S. Iyengar, A.I. Sarwat, O. Karabasoglu, A novel multi-time-scale modeling for electric power demand forecasting: from short-term to medium-term horizon. Electr. Power Syst. Res. **142**, 58–73 (2017)
42. M.H. Amini, A.I. Sarwat, S.S. Iyengar, I. Guvenc, Determination of the minimum-variance unbiased estimator for dc power-flow estimation, in *40th IEEE Industrial Electronics Conference (IECON 2014)* (Dallas, 2014)
43. P. Samadi, S. Bahrami, V.W.S. Wong, R. Schober, Power dispatch and load control with generation uncertainty, in *2015 IEEE Global Conference on Signal and Information Processing (GlobalSIP)*, Orlando, FL 2015, pp. 1126–1130
44. A. Ameli et al., A multiobjective particle swarm optimization for sizing and placement of DGs from DG owner's and distribution company's viewpoints. IEEE Trans. Power Delivery **29**(4), 1831–1840 (2014)
45. H. Kim, N. Feamster, Improving network management with software defined networking. IEEE Commun. Mag. **51**(2), 114–119 (2013)
46a. H. Arasteh, et al., SoS-based multiobjective distribution system expansion planning. Electr. Power Syst. Res. **141**, 392–406 (2016)
46b. S. Parhizi, A. Khodaei, Investigating the necessity of distribution markets in accommodating high penetration microgrids, in *IEEE PES Transmission & Distribution Conference & Exposition* (Dallas, 2016)
46c. H. Arasteh, M.S. Sepasian, V. Vahidinasab, An aggregated model for coordinated planning and reconfiguration of electric distribution networks. Energy **94**, 786–798 (2016)
47. Y. Yan, Y. Qian, H. Sharif, D. Tipper, A survey on cyber security for smart grid communications. IEEE Commun. Surv. Tutorials **14**(4), 998–1010 (2012)
48. M. Rahman, P. Bera, E. Al-Shaer, Smartanalyzer: a noninvasive security threat analyzer for AMI smart grid, in *Proceedings of IEEE INFOCOM*, Orlando, FL, 2012
49. M.H. Amini, M. Parsa Moghaddam, Probabilistic modelling of electric vehicles' parking lots charging demand, in *21th Iranian Conference on Electrical Engineering ICEE2013*, Ferdowsi University of Mashhad, 14–16 May 2013
50. A. Zidan, E.F. El-Saadany, A cooperative multi-agent framework for self-healing mechanisms in distribution systems. IEEE Trans. Smart Grid **3**(3), 1525–1539 (2012)
51. M.H. Amini, B. Nabi, M.-R. Haghifam, Load management using multi-agent systems in smart distribution network, in *Proceedings of IEEE Power and Energy Society General Meeting*, Vancouver, BC, July 2013, pp. 1–5
52. P. Dehghanian, Y. Wang, G. Gurrala, E. Moreno, M. Kezunovic, Flexible implementation of power system corrective topology control. Electr. Power Syst. Res. **128**, 79–89 (2015)
53. M. Kezunovic, T. Popovic, G. Gurrala, P. Dehghanian, A. Esmaeilian, M. Tasdighi, Reliable implementation of robust adaptive topology control, in *The 47th Hawaii International Conference on System Science, HICSS*, Big Island, 6–9 Jan 2014
54. P. Dehghanian, M. Kezunovic, Probabilistic impact of transmission line switching on power system operating states, in *IEEE Power and Energy Systems (PES) Transmission and Distribution (T&D) Conference and Exposition*, Dallas, 2–5 May 2016
55. P. Dehghanian, M. Kezunovic, Impact assessment of power system topology control on system reliability, in *IEEE Conference on Intelligent System Applications to Power Systems*, Porto, 11–16 Sept 2015

56. P. Dehghanian, M. Kezunovic, Probabilistic decision making for the bulk power system optimal topology control. IEEE Trans. Smart Grid **7**(4), 2071–2081 (2016)
57. Y. Shoham, K. Leyton-Brown, *Multi-agent Systems: Algorithmic.* Game Theoretic and Logical Foundations (Cambridge University Press, Cambridge, 2009–2010)
58. F. Bellifemine, G. Caire, D. Greenwood, *Developing Multi-Agent Systems with JADE* (Wiley, New York, 2007)
59. M. Wooldridge, G. Weiss, Intelligent agents, in *Multi-Agent Systems* (MIT Press, Cambridge, MA, 1999), pp. 3–51
60. P. Siano, C. Cecati, H. Yu, J. Kolbusz, Real time operation of smart grids via FCN networks and optimal power flow. IEEE Trans. Ind. Inf. **8**(4), 944–952 (2012)
61. A.I. Sarwat, M.H. Amini, A. Domijan Jr., A. Damnjanovic, F. Kaleem, Weather-based interruption prediction in the smart grid utilizing chronological data. J. Modern Power Syst. Clean Energy **4**(2), 308–315 (2016)
62. J.D. Glover, M.S. Sarma, *Power System Analysis and Design*, 3rd edn. (Brooks/Cole, Pacific Grove, CA, 2002)
63. B. Stott, Review of load-flow calculation methods. Proc. IEEE **62**, 916–929 (1974)
64. A.J. Wood, B.F. Wollenberg, *Power Generation, Operation and Control*, 2nd edn. (Wiley, New York, 1996)
65. G. Giannakis, V. Kekatos, N. Gatsis, S.-J. Kim, H. Zhu, B. Wollenberg, Monitoring and optimization for power grids: a signal processing perspective. IEEE Signal Process. Mag. **30**(5), 107–128 (2013)
66. L. Powell, DC load flow, Chap. 11, in *Power System Load Flow Analysis.* McGrawHill Professional Series (McGrawHill, New York, 2004)
67. B. Stott, J. Jardim, O. Alsac, DC power flow revisited. IEEE Trans. Power Syst. **24**(3), 1290–1300 (2009)
68. R.J. Kane, F.F. Wu, Flow approximations for steady-state security assessment. IEEE Trans. Circuits Syst. **CAS-31**(7), 623–636 (1984)
69. R. Baldick, Variation of distribution factors with loading. IEEE Trans. Power Syst. **18**(4), 1316–1323 (2003)
70. L. Xiong, W. Peng, L. Pohchiang, A hybrid AC/DC microgrid and its coordination control. IEEE Trans. Smart Grid **2**(2), 278–286 (2011)
71. S. Bera, S. Misra, P.C. Rodriguez Cloud computing applications for smart grid: a survey. IEEE Trans. Parallel Distrib. Syst. **99**, 1–18 (2014)
72. C.-T. Yang, W.-S. Chen, K.-L. Huang, J.-C. Liu, W.-H. Hsu, C.-H. Hsu, Implementation of smart power management and service system on cloud computing, in *Proceedings of IEEE International Conference on UIC/ATC*, 2012, pp. 924–929
73. S. Rusitschka, E. Kolja, C. Gerdes, Smart grid data cloud: a model for utilizing cloud computing in the smart grid domain, in *First IEEE International Conference on Smart Grid Communications (SmartGridComm)*, Gaithersburg, MD, 2010
74. M. Kezunovic, L. Xie, S. Grijalva, The role of big data in improving power system operation and protection, *IREP Symposium IEEE, Bulk Power System Dynamics and Control-IX Optimization, Security and Control of the Emerging Power Grid (IREP)*, 2013
75. C. Almond, A practical guide to cloud computing security, Aug 2009
76. B. Joshi, A. Santhana Vijayan, B. Kumar Joshi, Securing cloud computing environment against DDoS attacks, in *2012 International Conference on Computer Communication and Informatics (ICCCI-2012)*, Coimbatore, 10–12 Jan 2012
77. S. Yu, W. Zhou, R. Doss, Information theory based detection against network behavior mimicking DDoS attacks. IEEE Commun. Lett. **12**(4), 318–321 (2008)
78. Y. Shui, W. Zhou, Entropy-based collaborative detection of DDoS attacks on community networks, in *Sixth Annual IEEE International Conference on Pervasive Computing and Communications*, Piscataway, NJ, 2008, pp. 566–571
79. J.J.B. Krishnamurthy, M. Rabinovich, Flash crowds and denial of service attacks: characterization and implications for CDNs and web sites, in *Proceedings of International WWW Conferences*, 2002

80. Y. Chen, K. Hwang, Collaborative change detection of DDoS attacks on community and ISP networks, in *Proceedings of IEEE CTS*, 2006
81. R.B.G. Carl, G. Kesidis, S. Rai, Denial of service attack detection techniques. IEEE Internet Comput. **10**(1), 82–89 (2006)
82. B. Wang, Y. Zheng, W. Lou, Y.T. Hou, DDoS attack protection in the era of cloud computing and software-defined networking. Comput. Netw. **81**, 308–319 (2015)
83. A. Mishra, R. Mathur, S. Jain, J.S. Rathore, Cloud computing security. Int. J. Recent Innov. Trends Comput. Commun. **1**(1), 36–39 (2013)
84. A. Gupta, M.T. Hajiaghayi, H. Racke, Oblivious network design, in *SODA '06: Proceedings of the 17th Annual ACM-SIAM Symposium on Discrete Algorithm* (ACM, New York, 2006), pp. 970–979
85. R. Dewri, Location privacy and attacker knowl- edge: who are we fighting against? in *Security and Privacy in Communication Networks*. Lecture Notes of the Institute for Computer Sciences, Social Informatics and Telecommunications Engineering, vol. 96 (Springer, Berlin, 2012), pp. 96–115
86. M. Li, S. Salinas, A. Thapa, P. Li, *n*-CD: a geometric approach to preserving location privacy in location-based services, in *Proceedings of IEEE INFOCOM*, 2013
87. S.S. Iyengar, K.G. Boroojeni, N. Balakrishnan, *Mathematical Theories of Distributed Sensor Networks* (Springer, Berlin, 2014), pp. 111–145
88. R. Shi, M. Goswami, J. Gao, X.- feng Gu, Is random walk truly memory-less - traffic analysis and source location privacy under random walks, in *2013 Proceedings IEEE INFOCOM*, Turin, 2013, pp. 3021–3029
89. M. Gruteser, D. Grunwald, Anonymous usage of location-based services through spatial and temporal cloaking, in *ACM Mobisys'03*, May 2003
90. B. Gedik, L. Liu, Protecting location privacy with personalized k-anonymity: architecture and algorithms. IEEE Trans. Mob. Comput. **7**(1), 1–18 (2008)
91. J. Meyerowitz, R.R. Choudhury, Hiding stars with fireworks: location privacy through camouflage, in *Proceedings of ACM MobiCom*, Beijing, Sept 2009
92. M. F. Mokbel, C.Y. Chow, W.G. Aref, The new casper: query processing for location services without compromising privacy, in *Proceedings of VLDB*, 2006
93. P. Kalnis, G. Ghinita, K. Mouratidis, D. Papadias, Preventing location-based identity inference in anonymous spatial queries. IEEE Trans. Knowl. Data Eng. **19**(12), 1719–1733 (2007)
94. B. Gedik, L. Liu, Location privacy in mobile systems: a personalized anonymization model, in *Proceedings of IEEE ICDCS*, Columbus, OH, June 2005
95. C.-Y. Chow, M.F. Mokbel, X. Liu, A peer-to-peer spatial cloaking algorithm for anonymous location-based service, in *Proceedings of ACM GIS*, Arlington, VA, Nov 2006
96. A. Beresford, F. Stajano, Location privacy in pervasive computing. IEEE Pervasive Comput. **2**(1), 46–55 (2003)
97. B. Hoh, M. Gruteser, H. Xiong, A. Alrabady, Preserving privacy in GPS traces via uncertainty-aware path cloaking, in *Proceedings of ACM CCS 2007*, Alexandria, VA, Jan 2007
98. H. Kido, Y. Yanagisawa, T. Satoh, An anonymous communication technique using dummies for location-based services, in *Proceedings of IEEE ICPS*, Santorini, July 2006
99. H. Lu, C.S. Jensen, M.L. Yiu, Pad: privacy-area aware, dummy based location privacy in mobile services, in *Proceedings of ACM MobiDE*, Vancouver, June 2008
100. M. Duckham, L. Kulik, A formal model of obfuscation and negotiation for location privacy, in *Proceedings of International Conference on Pervasive Computing*, Munich, May 2005
101. C.A. Ardagna, M. Cremonini, S.D.C. di Vimercati, P. Samarati, An obfuscation-based approach for protecting location privacy. IEEE Trans. Dependable Secure Comput. **8**(1), 13–27 (2011)
102. A. Pingley, W. Yu, N. Zhang, X. Fu, W. Zhao, Cap: a contextaware privacy protection system for location-based services, in *Proceedings of IEEE ICDCS*, Montreal, June 2009
103. M. Damiani, E. Bertino, C. Silvestri, Probe: an obfuscation system for the protection of sensitive location information in LBS. Technical Report 2001-145, CERIAS, 2008

104. R. Shokri, G. Theodorakopoulos, C. Troncoso, J.-P. Hubaux, J.-Y. Le Boudec, Protecting location privacy: optimal strategy against localization attacks, in *CCS '12 Proceedings of the 2012 ACM Conference on Computer and Communications Security*, New York, NY, 2012, pp. 617–627
105. R. Shokri, G. Theodorakopoulos, J.-Y. Le Boudec, J.-P. Hubaux, Quantifying location privacy, in *2011 IEEE Symposium on Security and Privacy (SP)*, Berkeley, CA, May 2011, pp. 247–262
106. Z. Gong, G.-Z. Sun, X. Xie, protecting privacy in location-based services using K-anonymity without cloaked region, in *Eleventh International Conference on Mobile Data Management*, Kansas City, MO, 2010
107. R. Shokri, G. Theodorakopoulos, G. Danezis, J.-P. Hubaux, J.-Y. Le Boudec, Quantifying location privacy: the case of sporadic location exposure, in *Privacy Enhancing Technologies*, ed. by S. Fischer-Hubner, N. Hopper. Lecture Notes in Computer Science, vol. 6794 (Springer, Berlin, 2011), pp. 57–76

Part I
Physical Network Security

Chapter 2
Reliability in Smart Grids

This chapter introduces a reliable method of power distribution in a smart power network with high penetration of Distributed Renewable Resources (DRRs). From many reliability concerns regarding the smart grids, this chapter is devoted to the following couple of major issues: *power adequacy improvement* and *electric congestion prevention* in large-scale presence of green energy.

Section 2.1 provides an introduction to the literature of reliability in smart grids powered by conventional generation and green energy. Section 2.2 presents some preliminaries on how the reliability of smart grids can be measured utilizing the probabilistic methods and formal metrics. Afterwards, Sect. 2.3 proposes an example of reliability quantification of power networks with high penetration of distributed renewable (green) resources which have an increasingly important role in the future power networks.

Additionally, Sect. 2.4 addresses power congestion as a crucial reliability challenge in smart distribution power networks. It also introduces the mitigation mechanisms which help to reduce the power congestion in power lines of distribution networks. Finally, Sect. 2.5 summarizes the materials presented in this chapter in order to obtain an accurate understanding of some reliability issues in smart grids.

2.1 Introduction

Smart power grids deploy many technologies in order to turn the classic power network into a more distributed customer-driven one. On this way, electric vehicles [1], smart home appliances [2], Distributed Renewable Resources (DRRs) [3], and many other technologies have emerged and continue to emerge. One of the major challenges facing the aforementioned transition from a centralized generation-driven power network toward the modern smart power grids is balancing the demand and supply values which fluctuate much more than what is expected in a traditional

© Springer International Publishing Switzerland 2017
K.G. Boroojeni et al., *Smart Grids: Security and Privacy Issues*,
DOI 10.1007/978-3-319-45050-6_2

network [4, 5]. Without balancing the highly fluctuating values of demand and supply in the modern smart grids, the dream of enjoying its benefits will face major reliability issues as the loss of load probability will go through the roof if smart grids don't modernize its power resource management.

Here, we introduce a number of recent attempts to solve the aforementioned reliability issues in smart grids and keep the loss of load probability acceptably low in presence of DRRs and other new technologies which make the power demand values fairly unpredictable. Kekatos et al. presented an effective non-deterministic reactive power management scheme in [6]. Additionally, [7] proposed a distributed control scheme in order to manage the stochastic demand side of a smart grid by following the desired set-points for distributed level controllers. Likewise, another decentralized control approach was designed in [8] in order to have a low-cost DRR-based smart micro-grids. Moreover, [9] had a complete review of resource management schemes in a smart grid with non-deterministic nature of demand and generation.

In [10], Iyengar et al. proposed a new scheme for quantifiably reliable resource management and balancing the demand side and fluctuating generation of DRRs. Furthermore, they introduced a novel two-tier controlling scheme in [11] which handles the problem of uncertain demand and generation in modern architecture of smart power network. Additionally, distributed sensor networks, which are mathematically studied in [12], can help monitor the future power networks management and control.

A report of U.S. Department of Energy in 2008 [11] shows that recent progress in IT and communication technologies has made it possible to develop cost efficient power distribution systems. Smart distribution networks might have some specific functionalities such as self-healing, self-decision making, and resiliency [13, 14]. These requirements compel a new way of designing smart distribution grids for an economically-dispatched power distribution. As economic dispatch (ED) is one of the optimization problems related to the power systems operation, several techniques are used in the literature to solve this problem, e.g., mixed integer linear programming (MILP), constraint relaxation, and network approach [15]. Due to large-scale nature of the optimization problems in power systems, we have to utilize numerically efficient algorithms to overcome the complexity of these problems, specifically the ED problem [16].

Economic Dispatch ED is a major minimization problem in the field of power systems management and aims to compute the best way of satisfying the demand side of the power system by having enough committed generation units. In contrast, the unit commitment (UC) problem is another optimization problem with the goal of balancing the demand and generation of a power system on a day-ahead basis. This is while the ED is defined as the problem of balancing demand and generation for the next hour relying on the optimization solution of UC [17, 18].

Economic dispatch is considered as one of the most challenging issues facing the wide deployment of smarter grids with high penetration of DRRs [19]. A survey on the current methods of solving the ED problem has been accomplished in [9].

In the smart distribution networks, because of wide-usage of the DRRs, there is a need for designing new ED methods which prevents the power congestion in the connections and buses of a smarter network. Moreover, in demand side of the story, the customers have a more fluctuating usage pattern as a result of emergence of new technologies like electric vehicles, smart appliances, etc. [20, 21].

During many years of research and mathematical efforts to reach a solution for economic dispatch problem, many heuristic and strategies have been built to solve ED problem. In [22], an optimization strategy inspired by particle swarm is proposed to economically dispatch electricity in the network. The specialty of this optimization problem is that it is not solvable with conventional convex optimization strategies as the ED problem is not defined as a convex optimization problem. In [23], a multi-agent systems-based load management is introduced in a smart decentralized power network. In [24], the focus of addressing ED was placed on the structural form of the representing graph of the electric network topology.

Moreover, Sanjari et al. in [25] proposed a bilevel optimization scheme for economically dispatching the resources in a smart grid. This approach is a combination of two multi-scale optimization schemes (one daily and one hourly optimizer) with the aim of creating an online dispatcher. According to [26, 27], today's common ED solutions utilized MILP to find the best way of balancing demand and supply in a power network. Botterud et al. introduced a load dispatch when wind generation plays a significant role [28]. However, [29] considers the possibility of storing energy in storage units and turning the identity of ED optimization problem into a non-convex one which is not the same as the conventional convex problems. According to Li et al., there are some relaxation techniques that dissolve the non-convexity of the problem and make it possible for conventional methods to find the best dispatching solution [30] which minimizes the cost of dispatching power from the distributed generation toward the demand side. The objective function in this work constitutes two expressions, a deterministic cost of dispatching through the power connections and a stochastic power generation cost. Khator et al. proposed a mixed binary linear programming formulation to model the ED problem since the cost function can be modeled as a *concave* function of power flows.

In 2003, Yi Liu et al. in [21] introduced a minimization scheme for smart power dispatching in a power network which dispatches power from distributed generation through two-way power connections. In this work, which utilizes a pricing model with two levels, there is a high penetration of DRRs in the grid. In this model, the demand side is divided into two different power demands: deferrable and essential power demands. They modelled a scheme to satisfies both demand types by creating a trading policy and pricing protocol.

Oblivious Network Design One of the powerful mathematical tools that can play a crucial role in solving complex network optimization problems like ED is called oblivious network routing. There have been many efforts in the recent years by the researchers to develop various schemes for solving optimization problems with few known constraints. In some of these optimization problems, the objective function may change in different conditions and the constraints are not specified in advance.

In other words, oblivious network routing aims to give a solution to a set of optimization problems with a *set* of possible objective functions and a set of *possible* constraints. As a result, the solution to such problem does not aim to find the best solution for all the possible conditions. In fact, there may be no common solution to the set of given optimization problems. Instead, the oblivious network routing tries to give a routing solution to a set of network optimization problems so that the objective functions don't differ from their optimum values by an *approximation ratio* in all the cases [31, 32].

2.2 Preliminaries on Reliability Quantification

This section presents some preliminary concepts regarding the power adequacy, system reliability, and the quantification method for measuring the level of system reliability. Let $x_1, x_2, \ldots, x_n$ denote n customers spread over a residential smart power grid. Each customer x. has a power load profile $p.(t)$ which specifies the time path of the power load that the customer needs over time and expects the smart grid to satisfy it in a timely manner. Also, consider that this load profile consists of two terms: a deterministic predictable term $\hat{p}.(t)$ and a stochastic unpredictable term $\varepsilon.(t)$ which is considered as the *prediction error* of the load profile and assumed to be a Gaussian White Noise (GWN) of standard deviation less than σ_d.

On the generation side, there are two types of power generations that are distributed over the smart power grids and together manage to satisfy the customers' demand values over time: DRRs and conventional power plants. The conventional power plants as a global source of energy are expected to generate a deterministic *base generation* $G(t)$ over time. The DRRs, which are considered as the main source of power, are distributed over the power grid. Assume that $y_1, y_2, \ldots, y_n$ denote the n DRRs corresponding to the n customers and designed to satisfy their corresponding customers. Consider $g.(t)$ as the power generation profile of an arbitrary DRR $y.$. Also, let $\hat{g}.(t)$ and $\delta.(t)$, respectively, represent the predictable and stochastic part of the output for every renewable resource $y.$ where $\delta.(t)$ is a GWN of standard deviation less than σ_g.

Balancing Supply and Demand

After modeling the demand and generation side, it is time to come up with a scheme that aims to balance the supply and demand in a power network and make customers of the electric system satisfied. Based on the aforementioned model, the total generation load profile in the under-studied system is $G(t) + \sum_i \hat{g}_i(t) + \sum_i \delta_i(t)$, while its total demand load profile is $\sum_i \hat{p}_i(t) + \sum_i \varepsilon_i(t)$. The first step toward balancing the demand and supply is keeping the deterministic (predictable) part of demand profile satisfied by the deterministic part of the generation side; i.e.

$G(t)+\sum_i \hat{g}_i(t) \geq \sum_i \hat{p}_i(t)$. Additionally, considering the constraints of conventional power plants, the amount of power generation can't fall below some low-threshold G_{min}. As a result, we conclude that $G(t)$ should be set on the following value:

$$G(t) = \max \left\{ G_{min}, \sum_i \max\{0, \hat{p}_i(t) - \hat{g}_i(t)\} \right\}. \tag{2.1}$$

In the time periods that $G(t) = \sum_i \max\{0, \hat{p}_i(t) - \hat{g}_i(t)\}$, every customer receives a base generation of amount $g^{base}.(t) = \max\{0, \hat{p}.(t) - \hat{g}.(t)\}$. Otherwise, the generation base is used to charge the distributed storage units of the customers. Here, we continue the discussion in the following two cases:

Demand Exceeding the Supply

In this case, customers' demand exceeds the base generation plus the power generated by their corresponding DRRs; i.e. $p.(t) > g.(t) + \hat{p}.(t) - \hat{g}.(t)$ for a certain period of time. In such cases, the storage unit deployed in the customer side is charged with the rate of $\min\{C_{max}, \varepsilon.(t) - \delta.(t)\}$ over that specific period where C_{max} represents the maximum charging rate of the storage unit which is a physical characteristic of the unit.

Supply Exceeding the Demand

On the other hand, there are some cases that customers' demand falls below the power generated by their corresponding DRRs; i.e. $p.(t) < g.(t)$ for a certain period of time. In such cases, the storage unit deployed in the customer side is discharged with the rate of $\min\{D_{max}, g.(t) - d.(t)\}$ in that specific period where D_{max} represents the maximum discharging rate of the storage unit which is a physical characteristic of the unit.

2.3   System Adequacy Quantification

An electrical system maintains its adequacy as long as the customer demand is satisfied by the generation side of the system. Since we consider a storage unit connected to each customer, the system is called adequate if and only if the available energy $S.(t)$ to any customer $x.$ doesn't fall below a specific value s_{min}; i.e. $S.(t) \geq s_{min}$. Additionally, assume that the initial available energy to every customer is $S.(t) = s_0 > s_{min}$. Henceforth, the Loss of Load Probability for a given customer at the given moment t is defined as $1 - \mathbf{Pr}[\forall t' \leq t : S.(t') \geq s_{min}]$.

Here, we quantify the LOLP in the case that there exists no realtime change in the base generation, there is a high chance of having a power shortage. Mathematically, since $S.(t) = \int_0^t (g.(t') - p.(t') + g.^{(\text{base})}(t'))dt' + s_0$, the value of LOLP for a given customer x. would be not less than $\mathbf{Pr}\left[\sup_{t' \leq t}\{\int_0^t (\varepsilon.(t') - \delta.(t'))dt'\}\right] > s_0 - s_{\min}]$. Since $\varepsilon(t)$ and $\delta(t)$ are two GWN process, $\int_0^t (\varepsilon.(t') - \delta.(t'))dt'$ is a Brownian Motion Process and $\sup_{t' \leq t}\{\int_0^t (\varepsilon.(t') - \delta.(t'))dt'\}$ is a *running maximum process* which concludes that the LOLP value does not exceed $1 - \text{erf}\left(\frac{s_0 - s_{\min}}{\sqrt{2t(\sigma_d^2 + \sigma_g^2)}}\right)$.

2.4   Congestion Prevention: An Economic Dispatch Algorithm

This section discusses power congestion as one of the most important reliability challenges in smart distribution networks. Congestion prevention is a crucial constraint in dispatching power from generators to the demand side. Hence, the Economic Dispatch (ED) algorithms in smart grids need to prevent line power congestion while dispatching power efficiently. Although, this requirement is important in the classic deployment of centralized power generation in distribution networks, it gets more crucial in presence of distributed renewable resources where there are many paths for power flow to be dispatched to the demand side. As a result, congestion prevention in traditional power distribution networks turns into a challenging reliability issue in modern smart grids where there are a wide variety of power sources in the network and the topology of power lines is not *radial* anymore. This issue should be considered in the core of the ED algorithms where the power flows are routed through different paths in the complex topology of modern distribution networks. This section proposes a novel method of power dispatch which is based on a modern way of routing called *oblivious network routing*.

Oblivious routing is a heuristic way of approaching the Minimum-Cost Flow Problem (MCFP) which approximate the optimal congestion free solution to the flow problem when the current state of the network is unknown (oblivious). In fact, oblivious routing aims to find an acceptable solution of the MCFP when either the current traffic flow and/or flow cost functions are oblivious [31].

Consider a power distribution network which is designed to dispatch the power generated by the distributed generators $g_1, g_2, \ldots, g_n$ to a set of residential communities $c_1, c_2, \ldots, c_m$ through the network of communities $(\{c_i\}, E)$ where E is the set of power lines connecting two close by communities. The generation cost of the generator g_i for generating ρ units of power is a quadratic function of p with relatively small positive quadratic coefficient and large positive constant cost [33]. Also, assuming that power loss in the line $e \in E$ is a linear function of the power flow f_e, the traditional ED problem focuses on the problem of minimization the total generation cost subject to the following constraints:

- $f_e <$ line capacity of e for every $e \in E$.
- $m_i <$ generation of $g_i < M_i$ where m_i and M_i specify the low/high threshold of power generation for generator g_i.
- Kirchhoff's current laws.
- Total generation = total loss + total demand

In order to consider the congestion prevention challenge while solving the ED problem, we need to keep the value of fe's low while approximating the economically efficient solution. As [31] mathematically proves, the top-down oblivious network routing algorithm can solve the oblivious MCFP problem in a way that the total cost asymptotically competes with the optimal cost; while the line congestion does not exceed a constant fraction of its minimum possible congestion.

As another example of reliability-driven ED problem, consider a distribution network which meets $N - 1$ reliability criterion; i.e. the ED solution re-dispatches the power in the case of a failure in a power line. As an example, consider the IEEE 38-bus test system [36] in Fig. 2.1. This figure depicts one possible power dispatch configuration; which Fig. 2.2 shows another way of power dispatch if the line connecting busses 18 and 31 crashes. In such circumstances where the topology of the distribution network is changing and unknown, the economic dispatch algorithm can improve the process of power re-dispatching tremendously.

2.5 Summary and Conclusion

This chapter introduced a reliable method of power distribution in a smart power network with high penetration of Distributed Renewable Resources (DRRs). From many reliability concerns regarding the smart grids, this chapter was devoted to the following couple of major issues: *power adequacy improvement* and *electric congestion prevention* in large-scale presence of green energy. Section 2.1 provided an introduction to the literature of reliability in smart grids powered by conventional generation and green energy. Section 2.2 presented some preliminaries on how the reliability of smart grids can be measured utilizing the probabilistic methods and formal metrics. Afterwards, Sect. 2.3 proposed an example of reliability quantification of power networks with high penetration of distributed renewable (green) resources which have an increasingly important role in the future power networks.

Additionally, Sect. 2.4 addressed power congestion as a crucial reliability challenge in smart distribution power networks. It also introduced a mitigation mechanism based on oblivious network routing which helps to reduce the power congestion in power lines of distribution networks.

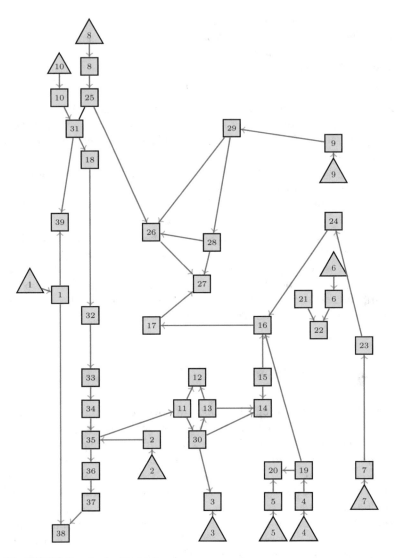

Fig. 2.1 IEEE 38-Bus standard test network

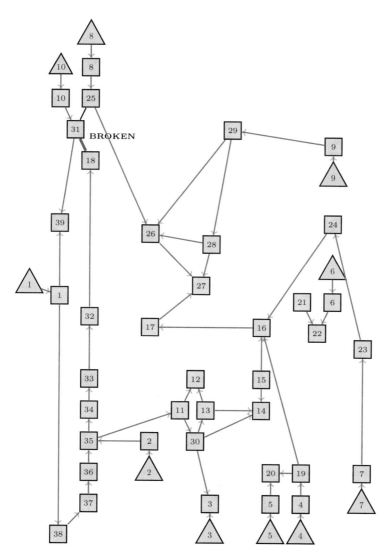

Fig. 2.2 IEEE 38-Bus standard test network

References

1. M.H. Amini et al., ARIMA-based demand forecasting method considering probabilistic model of electric vehicles parking lots, in *Proceedings of IEEE Power and Energy Society General Meeting*, 2015, pp. 1–5
2. M.H. Amini, J. Frye, M.D. Ilic, O. Karabasoglu, Smart residential energy scheduling utilizing two stage mixed integer linear programming, in *IEEE 47th North American Power Symposium (NAPS 2015)*, Charlotte, NC, 4–6 Oct 2015
3. A. Mohsenzadeh, M.-R. Haghifam, Simultaneous placement of conventional and renewable distributed generation using multi objective optimization, in *Proceedings of IEEE, Integration of Renewables into Distributed Grid Workshop, CIRED*, 2012
4. A. Papavasiliou, S.S. Oren, Supplying renewable energy to deferrable loads: algorithms and economic analysis, in *Proceedings of IEEE Power and Energy Society General Meeting*, Minneapolis, Minnesota, July 2010, pp. 1–8
5. A.R. Di Fazio, G. Fusco, M. Russo, Enhancing distribution networks to evolve toward smart grids: the voltage control problem, in *Proceedings of IEEE 52nd Annual Conference on Decision and Control (CDC)*, Firenze, Dec 2013, pp. 6940–6945
6. V. Kekatos, G. Wang, A.J. Conejo, G.B. Giannakis, Stochastic reactive power management in microgrids with renewables. IEEE Trans. Power Syst. **99**, 1–10 (2014)
7. S.E. Shafiei et al., A decentralized control method for direct smart grid control of refrigeration systems, in *Proceedings of IEEE 52nd Annual Conference on Decision and Control (CDC)*, Firenze, Dec 2013, pp. 6934–6939
8. G. Cavraro, R. Carli, S. Zampieri, A distributed control algorithm for the minimization of the power generation cost in smart micro-grid, in *Proceedings of IEEE 53rd Annual Conference on Decision and Control (CDC)*, Los Angeles, CA, Dec 2014, pp. 5642–5647
9. S. Kar, G. Hug, J. Mohammadi, J.M.F. Moura, Distributed state estimation and energy management in smart grids: a consensus+innovations approach . IEEE J. Sel. Top. Sign. Proces. **99**, 1–16 (2014)
10. S.S. Iyengar, K.G. Boroojeni, *Oblivious Network Routing Algorithms and Applications* (MIT Press, Cambridge, 2015)
11. U.S. Department of Energy, The smart grid: an introduction, 2008
12. S.S. Iyengar, K.G. Boroojeni, N. Balakrishnan, *Mathematical Theories of Distributed Sensor Networks* (Springer Publishing Company, Incorporated, Berlin, 2014)
13. A. Zidan, E.F. El-Saadany, A cooperative multi-agent framework for self-healing mechanisms in distribution systems . IEEE Trans. Smart Grid **3**(3), 1525–1539 (2012)
14. R.E. Brown, Impact of smart grid on distribution system design. in *Proceedings of IEEE Power and Energy Society General Meeting*, Pittsburgh, PA, July 2008, pp. 1–4
15. B.H. Chowdhury, S. Rahman, A review of recent advances in economic dispatch. IEEE Trans. Power Syst. **5**(4), 1248–1259 (1990)
16. V. Pappu, M. Carvalho, P. Pardalos, *Optimization and Security Challenges in Smart Power Grids* (Springer, Berlin, 2013)
17. J. Zhu, *Optimization of Power System Operation* (Wiley, New York, 2014)
18. N. Padhy, Unit commitment-a bibliographical survey. IEEE Trans. Power Syst. **19**(2), 1196–1205 (2004)
19. A. Papavasiliou, S.S. Oren, Supplying renewable energy to deferrable loads: algorithms and economic analysis, in *Proceedings of IEEE Power and Energy Society General Meeting*, Minneapolis, MN, July 2010, pp. 1–8
20. K.G. Boroojeni et al., Optimal two-tier forecasting power generation model in smart grids. Int. J. Inf. Process. **8**(4), 79–88 (2014)
21. Y. Liu, N. Ul Hassan, S. Huang, C. Yuen, Electricity cost minimization for a residential smart grid with distributed generation and bidirectional power transactions, in *IEEE Innovative Smart Grid Technologies (ISGT)*, Washington, DC, Feb 2013, pp. 1–6

22. A.I. Selvakumar, K. Thanushkodi, A new particle swarm optimization solution to nonconvex economic dispatch problems. IEEE Trans. Power Syst. **22**(1), 42–51 (2007)
23. M.H. Amini, B. Nabi, M.-R. Haghifam, Load management using multi-agent systems in smart distribution network, in *Proceedings of IEEE Power and Energy Society General Meeting*, Vancouver, BC, Canada, July 2013, pp. 1–5
24. J. Lavaei, D. Tse, B. Zhang, Geometry of power flows and optimization in distribution networks. IEEE Trans. Power Syst. **29**(2), 572–583 (2014)
25. M.J. Sanjari, H. Karami, H.B. Gooi, Micro-generation dispatch in a smart residential multi-carrier energy system considering demand forecast error. Energy Convers. Manag. **120**, 90–99 (2016)
26. E. Kellerer, F. Steinke, Scalable economic dispatch for smart distribution networks. IEEE Trans. Power Syst. **99**, 1–8 (2014)
27. M. Carrion, J. Arroyo, A computationally efficient mixed-integer linear formulation for the thermal unit commitment problem. IEEE Trans. Power Syst. **21**(3), 1371–1378 (2006)
28. A. Botterud, Z. Zhi, J. Wang et al., Demand dispatch and probabilistic wind power forecasting in unit commitment and economic dispatch: a case study of Illinois. IEEE Trans. Sustainable Energy **4**(1), 250–261 (2012)
29. Z. Li, Q. Guo, H. Sun, J. Wang, Sufficient conditions for exact relaxation of complementarity constraints for storage-concerned economic dispatch. IEEE Trans. Power Syst. **99**, 1–2 (2015)
30. S.K. Khator, L.C. Leung, Power distribution planning: a review of models and issues. IEEE Trans. Power Syst. **12**(3), 1151–1159 (1997)
31. S.S. Iyengar, K.G. Boroojeni, *Oblivious Network Routing: Algorithms and Applications* (MIT Press, Cambridge, 2015)
32. A. Gupta, M.T. Hajiaghayi, H. Racke, Oblivious network design, in *SODA '06: Proceedings of the 17th Annual ACM-SIAM Symposium on Discrete Algorithm* (ACM, New York, 2006), pp. 970–979
33. A.J. Wood, B.F. Wollenberg, *Power Generation, Operation, and Control* (Wiley, New York, 2012)
34. J. Fakcharoenphol, S.B. Rao, K. Talwar, A tight bound on approximating arbitrary metrics by tree metrics, in *Proceedings of the 35th STOC*, 2003, pp. 448–455
35. MATLAB version 8.5. Miami, Florida: The MathWorks Inc, 2015
36. University of Washington Electrical Engineering, Power systems test case archive (2015). Online Available: http://www.ee.washington.edu/research/pstca/
37. R.D. Zimmerman, C.E. Murillo-Sanchez, R.J. Thomas, MATPOWER: steady-state operations, planning and analysis tools for power systems research and education. IEEE Trans. Power Syst. **26**(1), 12–19 (2011)

Chapter 3
Error Detection of DC Power Flow Using State Estimation

In recent years, there is an ever-increasing concern about energy consumption and its environmental impacts, reliable energy supply, and sustainable development of energy and power networks. These issues motivate the evolution of Smart Grid (SG) as a novel means to worldwide electricity grid [1]. In this context, optimal operation of the power systems depends on finding the power flow through the transmission lines in the network. DC power flow has been widely used to tackle the power flow problem in the transmission networks. It is computationally efficient due to the linear nature of the problem. Smart grid includes several elements, such as electric vehicles [2–5], distributed renewable resources [6], smart appliances [7, 8], distributed state estimation and smart load management [9, 10], and demand response [11].

3.1 Introduction

In [3], a comprehensive method is proposed for the allocation of electric vehicle parking lots in the smart power distribution network. Furthermore, SG will gradually upgrade the power systems to achieve a more reliable, environmentally-friendly, and economically-operating system. A probabilistic model for electric vehicle parking lot modeling is proposed in [12]. This method is based on the probabilistic arrival and departure time and the expected driven distance of single electric vehicles. Self-healing and autonomous decision making methods play a pivotal role in the future power systems as complex dynamical networks [13–15]. Cloud computing applications attracted more attention for big data analytics, which plays a pivotal role in the issues related to the heterogeneous structure of the smart grids [16, 17]. Furthermore, big data analytics have been widely deployed to improve both power systems operation and protection [18]. In [18], three aspects of using large data sets in power systems are studied comprehensively: feature extraction, systematic

© Springer International Publishing Switzerland 2017
K.G. Boroojeni et al., *Smart Grids: Security and Privacy Issues*,
DOI 10.1007/978-3-319-45050-6_3

integration for power system applications, and examples of typical applications in the utility industry. A comprehensive study of the smart grid effects on the smart distribution network design is provided in [14]. In [15], multi-agent systems are utilized to develop a smart load management strategy for the smart distribution networks. According to [15], multi-agent systems include multiple autonomous agents (including software and hardware) with diverging information or interest [19]. These systems are widely used in different applications, from small scale systems which are used for personal purposes, to complex large-scale applications for industrial purposes [20]. According to [15, 21] an agent has the following features:

- Autonomy: ability to control its actions and internal state without the direct intervention of human or other agents;
- Social-ability: agents can interact with human and other agents utilizing agent communication language;
- Reactivity: agents can analyze the environment and respond to the corresponding changes that happens in it;
- Pro-activeness: the response of agents to the environment is in object-oriented direction.

Optimal operation of microgrids and the smart grids is obtained by developing an energy management system in [22]. The proposed method is based on hybrid connected neural networks and optimal power flow.

Power flow problem is solved to obtain the steady-state operation point of the power systems. In the optimal power flow problem, contingencies (fault, congestion, physical attack, generator failure) and physical limits of network are considered as the constraints while solving the power flow problem. Furthermore, the contingency analysis plays an important role in the smart grid. In [23], an interruption prediction approach is proposed to improve the reliability of the smart grid using the historical weather data and the chronological interruption data. To this end, Sarwat et al. used artificial neural networks to perform an accurate forecast based on the available historical data and the effectiveness of their proposed method is validated via real-world historical data from Florida Power and Light (FPL) as well as National Climatic Data Center (NCDC). Furthermore, the real-time contingency data can be used to prevent the issues caused by outages. Smart grids require a robust and secure self-decision making ability. To this end, power flow calculations should be performed accurately. Any computation errors due to unintentional faults or bad data injection to the system in the power flow calculation may lead to a major outage or black-out [24].

The ultimate objective of solving power flow problem is to determine the amount of voltage phase and voltage angles, active power, and reactive power in the power system. Four widely used methods in the power flow analysis can be listed as [25, 26]:

- **Gauss–Seidal method** is based on an iterative method for solving linear systems. This approach can be applied to the problems with nonsingular matrix representation. Note that nonsingular matrix refers to a matrix which its determinant is nonzero. It is also called full-rank matrix. The matrix should be diagonally dominant or symmetric and positive definite so that the convergence of method is guaranteed.

- **Newton Raphson method** is another technique for solving power flow problem. The convergence of this method is quadratic. However this method is used in the literature for power flow studies, there are some issues with the method.
- **Fast Decoupled method**: Fast power flow algorithms are uses to expedite the power flow computation into short durations (seconds or a proportion of seconds).
- **DC power flow method**[27]: Further simplifications such as neglecting the $Q-V$ equation and assuming the voltage magnitudes are constant at 1.0 per-unit will lead to DC power flow formulation [28]. This simplification and the equation related to the DC power flow will be elaborately discussed in a separate section. In [29], some of the advantages of DC power flow are mentioned as follow

  (*i*) Non-iterative, reliables and unique solutions;
  (*ii*) Simplified methods and software;
  (*iii*) Efficient solution and optimization, specifically for contingency analysis;
  (*iv*) Minimal and easy-to-obtain network data requirement;
  (*v*) Linear structure that matches the economic theory for transmission-oriented market studies;
  (*vi*) Fairly accurate approximation of active power flows.

  Some studies tried to analyze the theoretical error of DC models in power flow [30, 31]. Hybrid AC/DC microgrids control requires DC power-flow method for studying the technical aspects [32].

3.2 Preliminaries of the DC Power Flow and State Estimation

Assume that the transmission line between the ith and jth buses, shown in Fig. 3.1, is represented by its impedance $Z_{ij} = R_{ij} + jX_{ij}$, where R_{ij} and X_{ij} represent the Resistance and Reactance of line ij, respectively.

We can calculate the admittance of line ij as $Y_{ij} = 1/Z_{ij} = g_{ij}+jb_{ij}$, where g_{ij} and b_{ij} represent the Conductance and Susceptance of line ij, respectively. Afterwards, the active power flow (P_{ij}) and reactive power flow (Q_{ij}) of this line are calculated using the following equation [33],

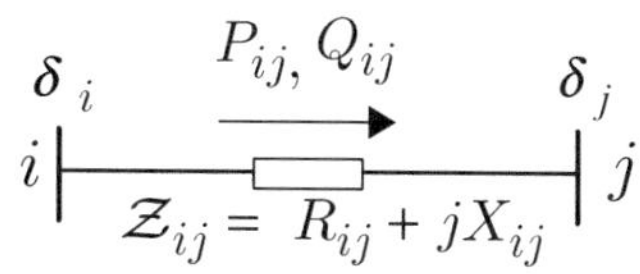

Fig. 3.1 Power flow between bus i and j

$$P_{ij} = -V_i^2 g_{ij} + V_i V_j g_{ij} \cos(\delta_i - \delta_j) + V_i V_j b_{ij} \sin(\delta_i - \delta_j),$$

$$Q_{ij} = -V_i^2 b_{ij} + V_i V_j g_{ij} \sin(\delta_i - \delta_j) - V_i V_j b_{ij} \cos(\delta_i - \delta_j),$$

where δ_i and V_i represent the voltage angle and voltage magnitude at the ith bus, respectively. Furthermore, we can calculate the total active and reactive power injections at bus k using the following equations:

$$P_k = \sum_{j=1}^{N} |V_k||V_j|(g_{kj} \cos(\delta_j - \delta_k) + b_{kj} \sin(\delta_j - \delta_k))$$

$$Q_k = \sum_{j=1}^{N} |V_k||V_j|(g_{kj} \sin(\delta_j - \delta_k) - b_{kj} \cos(\delta_j - \delta_k)).$$

As we mentioned, $\mathcal{Y}_{ij} = 1/\mathcal{Z}_{ij} = g_{ij} + jb_{ij}$; by substituting $\mathcal{Z}_{ij}$, we have

$$\mathcal{Y}_{ij} = 1/\mathcal{Z}_{ij} = \frac{1}{R_{ij} + jX_{ij}} = \frac{R_{ij}}{R_{ij}^2 + X_{ij}^2} - j\frac{X_{ij}}{R_{ij}^2 + X_{ij}^2}.$$

Consequently, we have $g_{ij} = \frac{R_{ij}}{R_{ij}^2 + X_{ij}^2}$ and $b_{ij} = -\frac{X_{ij}}{R_{ij}^2 + X_{ij}^2}$. If R_{ij} is very small compared to X_{ij}, then we will g_{ij} will be very small compared to b_{ij}.

The DC power flow model is based on the following three assumptions.

1. In the high voltage transmission line, the resistive part of the impedance can be neglected because of the large value of X_{ij}/R_{ij} [27]. Therefore, the network is fairly inductive, i.e. $R_{ij} = 0$ which leads to $g_{ij} = 0$ and $b_{ij} = -\frac{1}{X_{ij}}$.
2. We can neglect the voltage angle difference of any connected buses ($\delta_i - \delta_j \approx 0$). Hence, we can approximate the trigonometric functions as $\cos(\delta_i - \delta_j) \approx 1$ and $\sin(\delta_i - \delta_j) \approx 0$.
3. Voltage magnitude at all buses of the power network is assumed to be one per unit. This assumption is feasible due to the normal operation conditions of the power system.

The mentioned assumptions can be summarized as

$$\begin{cases} |V_i| = 1 \\ R_{ij} = 0 \rightarrow B_{ij} = \frac{1}{X_{ij}} \\ \cos(\delta_i - \delta_j) \approx 1 \\ \sin(\delta_i - \delta_j) \approx 0 \end{cases} ; \forall i, j \in \mathcal{T} \tag{3.1}$$

where $\mathcal{T}$ denotes the set of all buses. Considering these assumptions and Fig. 3.1, active power can be calculated using voltage phase values between buses i and j as shown in (3.2).

$$P_{ij} = \frac{\delta_i - \delta_j}{X_{ij}},$$
(3.2)

We define $\mathbf{p}$ as the vector of active power values and δ as the vector of voltage phases. We define $\mathbf{p} = \mathbf{B}\delta$, where the entries of $\mathbf{B}$ are obtained by

$$\mathbf{B}_{ij} = \begin{cases} \sum_{k \in \mathcal{T} \setminus \{i\}} X_{ik}^{-1} & ; i = j, \ \forall i \in \mathcal{T} \\ -X_{ij}^{-1} & ; i \neq j, \ \text{line } ij \text{ exists}, i, j \in \mathcal{T} \\ 0 & ; \text{otherwise} \end{cases}$$

3.2.1   Introduction to State Estimation

Estimation theory plays a pivotal role in many electrical engineering, electronic systems, power systems, and signal processing applications which are designed to extract the information [34]. Here we list some of the major applications of estimation theory with emphasis on future power systems, referred to as smart grids.

- **Energy Management Systems in the Smart Grids** State estimation is one of the crucial means in Energy Management Systems (EMS) in the smart grids [35]. Measurement equipments send their information using the Supervisory Control and Data Acquisition (SCADA) system to the decision and control center. Smart meters, Phasor Measurement Units (PMUs), and other sensors measure the system parameters such as voltage magnitudes and angles and power injections at nodes or power flows and currents on lines [10]. A state estimator transforms measured information about the systems parameters into an estimate of the true state of the system and aims to detect and eliminate the measurement noise [36]. According to [37], measurements can include both analog and digital information. The Remote Terminal Units are located at substations and collect several types of information. The data then is sent to the control center. In recent years, Intelligent Electronic Devices (IEDs) are being used as an effective alternative for RTUs. The combination of these RTUs and IEDs, as well as Local Area Network (LAN) collects the data and sends it to SCADA as a means for communication with the control center for further processing. Moreover, we can include the modern Advanced Metering Infrastructure (AMI) as a means of collecting end-user data for EMS purposes. The measurements can include several parameters such as power flow of the lines, bus voltage magnitude and voltage angle, and electric load demand. EMS applications cover a variety of functions such as automatic generation control, contingency analysis[38], congestion management, load forecasting [39], demand response [11], and load management [15]. In [38], the estimation of post-contingency voltages and reactive power generation and flows is investigated by deploying the sensitivities.

To this end, piece-wise linear estimates are employed to evaluate the effect of power system equipment on the obtained estimates.

- **DC Power Flow Estimation** In our recent studies, we studied the application of state estimation in DC power flow estimation. In [40], we introduced a comprehensive formulation for determination of the minimum variance unbiased estimators (MVUEs) to calculate the active power flow based on the voltage phase at different buses of the analyzed power system. According to [41], a new DC power flow estimation approach is developed for smart grid estimation purposes. The sequential linear minimum mean square error (LMMSE) estimator was deployed to solve the DC power flow problem [41]. We introduced an accurate approach for measurement error detection in DC power flow problem in [42].
- **Speech Recognition** In [43], the fundamentals of speech recognition using computers are introduced. In a related context, [44] developed an approach to estimate the parameters of hidden Markov models of speech. Digalakis et al. in [45] introduced a nontraditional method for parameter estimation of a stochastic linear system. Their approach built on the expectation-maximization algorithm which can be behaved as the continuous analog of the Baum–Welch estimation algorithm for hidden Markov models[45].
- **Channel Estimation in Wireless Communication** In [46], the effect of channel estimation error on the performance of wireless communication networks is theoretically studied. In [47], a low-rank channel estimators for orthogonal frequency-division multiplexing (OFDM) systems is developed. According to [48], robust channel estimation is investigated for OFDM systems by deriving a minimum mean-square-error channel estimator, which is not sensitive to the channel statistics.

3.3   Minimum-Variance Unbiased Estimator (MVUE)

In order to improve robustness of the future power systems against bad data injection and measurement error, accurate estimation of the power flow problem is required. According to [37], one of the main reasons of New York power outage at 1987 was erroneous information exchange about the status of power networks. Moreover, Abur and Exposito described state estimation as an inevitable part of market operation in the new markets such as PJM. Hence, we need novel estimation methods to upgrade the conventional approaches to solve DC power flow problem. In this section, we provide a comprehensive formulation for minimum variance unbiased estimators (MVUEs) to find the active power flow based on the voltage phase at different buses of the analyzed power system. The state variables are the voltage phases and the received measurement signals are active power measurements. We developed this method in [40].

3.3.1   Measurement Error Representation in the Linear DC Power Flow Equation

The following equation illustrates the linearized form of power flow considering the measurement errors:

$$\mathbf{x} = \mathbf{H}\boldsymbol{\delta} + \mathbf{e}, \tag{3.3}$$

where $\mathbf{x}$ denotes the active power measurements vector, $\mathbf{H}$ represents a square matrix which is determined based on the topology of the power systems and the inductance values. $\mathbf{x}$, $\mathbf{H}$, $\boldsymbol{\delta}$, and $\mathbf{e}$ can be represented as follows:

$$\mathbf{x} = [P_1 \ P_2 \ \dots \ P_n]^T, \tag{3.4}$$

$$\boldsymbol{\delta} = [\delta_1 \ \delta_2 \ \dots \ \delta_n]^T, \tag{3.5}$$

$$\mathbf{e} \sim N(\mathbf{0}, \mathbf{R}), \tag{3.6}$$

$$H_{\alpha\beta} = \begin{cases} H_{\alpha\beta} = \sum_{i=1, i\neq\alpha}^{N} B_{\alpha i} & ; \alpha = \beta \\ -B_{\alpha\beta} & \text{otherwise} \end{cases}. \tag{3.7}$$

In order to obtain the appropriate estimator, we consider the noise is to be colored Gaussian noise. Hence, the joint probability density function (pdf) of $\mathbf{x}$ is represented as

$$p(\mathbf{x}; \boldsymbol{\delta}) = \frac{1}{(2\pi)^{n/2} |\mathbf{R}|^{1/2}} \exp\left\{ \frac{-1}{2}(\mathbf{x} - \mathbf{H}\boldsymbol{\delta})^T \mathbf{R}^{-1}(\mathbf{x} - \mathbf{H}\boldsymbol{\delta}) \right\}, \tag{3.8}$$

where $|\mathbf{R}| = \det(\mathbf{R})$ shows the determinant of square matrix R. Here, we defined the preliminaries required for problem formulation. We introduce the generalized linear model which is necessary for obtaining the minimum-variance unbiased estimator (MVUE) for DC power flow state estimation problem. Hereafter, the linear models will be introduced for a specific noise vector. Then, the general linear model is discussed [34, 40].

3.3.2   Linear Model

Considering a noise vector with PDF let $\mathcal{N}(\mathbf{0}, \sigma^2\mathbf{I})$ denote the pdf for a noise vector. $\hat{\boldsymbol{\delta}} = \mathbf{g}(\mathbf{x})$ represents a minimum-variance unbiased estimator (MVUE) if:

$$\frac{\partial \ln p(\mathbf{x}; \boldsymbol{\delta})}{\partial \boldsymbol{\delta}} = \mathbf{I}(\boldsymbol{\delta})(\mathbf{g}(\mathbf{x}) - \boldsymbol{\delta}), \tag{3.9}$$

where $C_{\hat{\delta}} = I^{-1}(\delta)$. As a result, we should factor the derivative of the natural logarithm of (8) into the form $\mathbf{I}(\delta)(\mathbf{g}(\mathbf{x}) - \delta)$ in order to obtain the MVUE.

$$\frac{\partial \ln p(\mathbf{x}; \delta)}{\partial \delta} = \frac{\partial}{\partial \delta} \left[-\ln(2\pi\sigma^2)^{N/2} - \frac{1}{2\sigma^2} (x - \mathbf{H}\delta)^T (x - \mathbf{H}\delta) \right]. \qquad (3.10)$$

We can determine the MVUE for δ after performing some manipulations. The MVUE can be shown as the following equation:

$$\hat{\delta} = (\mathbf{H}^T \mathbf{R}^{-1} \mathbf{H})^{-1} \mathbf{H}^T \mathbf{R}^{-1} \mathbf{x}, \qquad (3.11)$$

Note that the covariance matrix is given by $\mathbf{C}_{\hat{\delta}} = \sigma^2 (\mathbf{H}^T \mathbf{H})^{-1}$.

3.3.3 Generalized Linear Model for State Estimation

To obtain the generalized model for DC power flow state estimation, we need to satisfy two extensions. This helps us to obtain the generalized model that potentially can be utilized for general Gaussian noise rather than white Gaussian noise. Let $\mathbf{x}$ denote the vector that may contain a known signal term. We denote the known part of the signal by $\mathbf{s}$. Hence we can write $\mathbf{x}$ as [34]:

$$\mathbf{x} = \mathbf{H}\delta + \mathbf{s} + \mathbf{e}, \qquad (3.12)$$

where $\mathbf{s}$ is a vector of known signal samples with the same length as $\mathbf{x}$, and $\mathbf{e}$ denotes the noise vector with general Gaussian pdf $\mathcal{N}(\mathbf{0}, \mathbf{R})$. For the sake of simplification in the modeling procedure, we can use the whitening transformation. Whitening transformation helps us to transform the problem to the desired linear model which is described in this section. If we represent the noise covariance matrix, $\mathbf{R}$, as:

$$\mathbf{R}^{-1} = \mathbf{D}^T \mathbf{D}, \qquad (3.13)$$

then for the whitening transformation purpose, we can use the matrix $\mathbf{D}$. If $\mathbf{e}' = \mathbf{D}\mathbf{e}$, then pdf of $\mathbf{e}'$ is $\mathcal{N}(\mathbf{0}, \mathbf{I})$. After performing this transformation we have

$$\mathbf{x}' = \mathbf{D}\mathbf{x} = \mathbf{D}\mathbf{H}\delta + \mathbf{D}\mathbf{s} + \mathbf{D}\mathbf{e} = \mathbf{H}'\delta + \mathbf{s}' + \mathbf{e}'. \qquad (3.14)$$

Now, we define $\mathbf{x}'' = \mathbf{x}' - \mathbf{s}'$. So we have

$$\mathbf{x}'' = \mathbf{H}'\delta + \mathbf{w}'$$

The MVUE of δ for the given observed data $\mathbf{x}''$ is

$$\hat{\delta} = (\mathbf{H}'^T \mathbf{H}')^{-1} \mathbf{H}'^T \mathbf{x}'' = (\mathbf{H}^T \mathbf{D}^T \mathbf{D}\mathbf{H})^{-1} \mathbf{H}^T \mathbf{D}^T \mathbf{D}(\mathbf{x} - \mathbf{s}). \qquad (3.15)$$

By replacing $\mathbf{D}^T \mathbf{D}$ with $\mathbf{R}^{-1}$, we have

$$\hat{\delta} = (\mathbf{H}^T \mathbf{R}^{-1} \mathbf{H})^{-1} \mathbf{H}^T \mathbf{R}^{-1} (\mathbf{x} - \mathbf{s}), \qquad (3.16)$$

and the covariance matrix is shown by $\mathbf{C}_{\hat{\delta}} = (\mathbf{H}^T \mathbf{R}^{-1} \mathbf{H})^{-1}$. Now, the problem is defined and the tools for solving linear model and calculate the MVUE for a linear model for the specific noise (white Gaussian) and general noise (general Gaussian) is discussed elaborately.

The detailed case study to prove the effectiveness of the obtained MVUE is provided in our study in [40]. In order to investigate the effect of noise covariance matrix on the accuracy of estimation process, we defined three different scenarios. The results of our study showed that higher sparsity in the covariance matrix (less correlation between noise vector elements) leads to better accuracy [40]. Moreover, in the case in which the noise is uncorrelated, the estimation results are more accurate than the case in which the noise is correlated.

3.4 Bayesian-Based LMMSE Estimator for DC Power Flow Estimation

In this section, we use the linear model from (3.3) and develop a sequential linear minimum mean square error (LMMSE) estimator for solving the DC power flow state estimation. Accurate state estimation approaches are required to obtain an acceptable adequacy level for the complex smart grid studies. We introduce the classic linear estimator model with deterministic inputs. Then, we explain the detailed formulation of the LMMSE estimator with random variables as input. Note that the prior probability density function (pdf) of these random variables is known.

In [41], a novel DC power flow estimation approach is introduced. This approach is applicable to smart grids for the estimation purposes based on the information from wide area measurement systems. The sequential linear minimum mean square error (LMMSE) estimator was introduced to solve the DC power flow problem.

3.4.1 Linear Model

We use the introduced linearized form of power flow considering the measurement errors as $\mathbf{x} = \mathbf{H}\boldsymbol{\delta} + \mathbf{e}$. As it has been mentioned, $\mathbf{x}$ includes the active power measurements, $\mathbf{H}$ is a square matrix. Hereafter, we will use $\mathbf{x} = \mathbf{B}\boldsymbol{\delta} + \mathbf{e}$ alternatively as the linear model representation. Additionally, we assume that the noise is colored Gaussian. The joint pdf of $\mathbf{x}$ is shown in (3.17)

$$p(\mathbf{x}; \boldsymbol{\delta}) = \frac{1}{(2\pi)^{n/2} |\mathbf{R}|^{1/2}} \exp\left\{ \frac{-1}{2} (\mathbf{x} - \mathbf{B}\boldsymbol{\delta})^T \mathbf{R}^{-1} (\mathbf{x} - \mathbf{B}\boldsymbol{\delta}) \right\}, \qquad (3.17)$$

where $|\mathbf{R}|$ denotes the determinant of a matrix $\mathbf{R}$.

Consequently, the linear model of DC power flow is extracted and explained. In the next subsection, Maximum Likelihood Estimator **MLE** and **LMMSE** estimator are discussed.

3.4.2 Bayesian Linear Model

Consider the Bayesian Linear model in $\mathbf{x} = \mathbf{B}\delta + \omega$ form, where δ is the to-be-estimated parameter with *prior pdf* $\mathcal{N}(\mu_\delta, \mathbf{R}_\delta)$ and ω is a white Gaussian noise with *pdf* $\mathcal{N}(0, \mathbf{R}_\omega)$ [34]. Therefore we have

$$\begin{cases} E(\mathbf{x}) = \mathbf{B}\mu_\delta \\ E(\delta) = \mu_\delta \end{cases} \tag{3.18}$$

As $\hat{\delta} = E(\delta|\mathbf{x})$ and $E(\delta|\mathbf{x}) = E(\delta) + \mathbf{R}_{\delta x}R_{xx}(\mathbf{x} - E(\mathbf{x}))$, considering (3.18) we can obtain the estimator as shown in (3.19).

$$\hat{\delta} = \mu_\delta + \mathbf{R}_\delta\mathbf{B}^T(\mathbf{B}\mathbf{R}_\delta\mathbf{B}^T + \mathbf{R}_\omega)^{-1}(\mathbf{x} - \mathbf{B}\mu_\delta) \tag{3.19}$$

3.4.3 Maximum Likelihood Estimator for DC Power Flow Estimation

In order to find *MLE* of δ, we differentiate the log-likelihood from (3.17), which is represented in (3.20).

$$\frac{\partial \ln p(\mathbf{x}; \delta)}{\partial \delta} = \frac{\partial(\mathbf{B}\delta)^T}{\partial \delta}\mathbf{R}^{-1}(\mathbf{x} - \mathbf{B}\delta) \tag{3.20}$$

After some manipulations and simplifications, the MLE of δ is extracted and shown in (3.21).

$$\hat{\delta} = (\mathbf{H}^T\mathbf{R}^{-1}\mathbf{H})^{-1}\mathbf{H}^T\mathbf{R}^{-1}\mathbf{x}, \tag{3.21}$$

In MLE approach, however the value of δ is unknown, it is deterministic. Therefore the estimated value, $\hat{\delta}$, is independent of the actual values of δ.

3.4.4 Bayesian-Based Linear Estimator for DC Power Flow

In the Bayesian linear model, we consider δ as a random variable with a specific prior *pdf* denoted by $p(\delta)$. This prior information will provide more accurate estimator than the Maximum Likelihood Estimator for the introduced linear model.

Assuming $p(\mathbf{x}, \delta)$ is the joint *pdf* of $\mathbf{x}$ and δ. In the Bayesian method, we can derive the minimum mean square error (MMSE) estimator by minimizing $\hat{\delta} = \arg\min_{\hat{\delta}} B_{\text{MSE}}(\hat{\delta})$. Bayesian mean square error (B_{MSE}) is represented in (3.22).

$$B_{\text{MSE}}(\hat{\delta}) = E[(\delta - \hat{\delta})^2] = \int \int (\delta - \hat{\delta})^2 p(\mathbf{x}, \delta)\, d\mathbf{x}\, d\delta \tag{3.22}$$

To obtain the MMSE we should differentiate B_{MSE} with respect to $\hat{\delta}$ and set it to zero. The estimator and *posterior pdf* are given in (3.23) and (3.24), respectively.

$$\hat{\delta} = E(\delta|\mathbf{x}) = \int \delta p(\delta|\mathbf{x}) \tag{3.23}$$

$$p(\delta|\mathbf{x}) = \frac{p(\mathbf{x}|\delta)p(\delta)}{\int p(\mathbf{x}|\delta)p(\delta)\, d\delta}. \tag{3.24}$$

Bayesian Gauss Markov Theorem [34] For the Bayesian linear model form, $\mathbf{x} = \mathbf{B}\delta + \omega$ (the pdf of δ and prior pdf of noise are defined in previous section), then the *LMMSE* estimator of δ is

$$\hat{\delta} = E(\delta) + (\mathbf{R}_{\delta\delta}^{-1} + \mathbf{B}^T\mathbf{R}_{\omega}^{-1}\mathbf{B})^{-1}\mathbf{B}^T\mathbf{R}_{\omega}^{-1}(\mathbf{x} - \mathbf{B}E(\delta)). \tag{3.25}$$

$\epsilon = \delta - \hat{\delta}$ denote the error whose mean is zero and its covariance matrix is provided in (3.26).

$$\mathbf{R}_{\epsilon} = (\mathbf{R}_{\delta\delta}^{-1} + \mathbf{B}^T\mathbf{R}_{\omega}^{-1}\mathbf{B})^{-1}. \tag{3.26}$$

This error value can be used to evaluate the performance of the estimator.

3.4.5   Recursive Bayesian-Based DC Power Flow Estimation Approach for DC Power Flow Estimation

The goal is to estimate the voltage angle in a recursive manner. Let $\hat{\delta}[n + 1]$ denote the voltage angle, which is $\hat{\delta}$ based on $n + 1$ samples of data. Also let $\hat{\delta}[n]$ represent the estimation based on n previous data samples, and $\mathbf{x}[n + 1]$, the $(n + 1)^{th}$ data sample [41]. $\hat{\mathbf{x}}[n + 1|n]$ is the estimation value of $\mathbf{x}[n + 1]$ using previous n data samples. Using this approach, we can estimate $\hat{\delta}[n + 1]$ based on the available data samples [34, 49]:

1. Based on *Gauss–Markov theorem* we estimate $\hat{\delta}[n]$, or $\hat{\delta}[n] = \hat{\delta}[0]$, as an initial estimation. Here $\hat{\delta}[n]$ is considered as the real value of voltage angle projected on the subspace spanned by $\{\mathbf{x}[0], \mathbf{x}[1], \ldots, \mathbf{x}[n]\}$.

2. In this step the LMMSE estimator of $x[n + 1]$ is obtained based on the available n samples, or we consider initial estimator, $\hat{x}[1|0]$. $\hat{x}[n + 1|n]$ can be used as the true value $x[n + 1]$ projected to the subspace spanned by $\{x[0], x[1], \ldots, x[n]\}$.
3. *Innovation* is defined by $x[n] - \hat{x}[n + 1|n]$ which is orthogonal to the subspace spanned by $\{x[0], x[1], \ldots, x[n]\}$. Innovation serves as the direction for obtaining new sample $x[n + 1]$.
4. In order to calculate the *correction gain*, $K[n+1]$, we normalize the projection of δ on the innovation $x[n] - \hat{x}[n+1|n]$. Correction gain equation is shown in (3.27).

$$K[n + 1] = \frac{E[\delta(x[n] - \hat{x}[n + 1|n])]}{E[(x[n] - \hat{x}[n + 1|n])^2]} \tag{3.27}$$

5. Eventually, the estimator is updated by adding previous estimator to the correction as represented in (3.28).

$$\hat{\delta}[n + 1] = \hat{\delta}[n] + K[n](x[n] - \hat{x}[n + 1|n]) \tag{3.28}$$

Let $M[n]$ denote the error covariance matrix, which is derived based on (3.29).

$$M[n] = E[(\delta - \hat{\delta}[n + 1])(\delta - \hat{\delta}[n + 1])^T] \tag{3.29}$$

3.5 Error Detection Using Sparse Vector Recovery

In this section we introduce a novel approach which can be used for detecting the measurement error accurately in the DC power flow state estimation problem. Our proposed approach builds on the singularity of the impedance matrix in DC power flow problem. It also takes advantage of the sparsity of the error vector and solves the DC power flow problem as a sparse vector recovery problem. This approach utilizes l_1-norm minimization for state estimation. It has been shown that this method computes the sparse measurement errors exactly. Furthermore, its performance is not deteriorated by the magnitudes of the measurement errors. Consequently, the proposed novel error detection method for DC power flow estimation can identify the noisy elements if the measurements include an additive white Gaussian noise (AWGN) plus sparse noise with large magnitude. The source of this noise can be either measurement error or bad data injection attacks.

We studied this approach in [42] for the first time. We introduced an accurate approach for measurement error detection in DC power flow problem. First, we transformed the DC power flow problem into a sparse vector recovery problem. Afterwards, a l_1-norm minimization approach was proposed which is capable of exactly calculating the measurement errors.

3.5.1   Sparse Vector Recovery

Many signal/data of interests follow some low dimensional structures which are usually ignored by the classic data analysis methods. Recently, Rahmani and Atia have focused on developing new approaches to leverage these low dimensional structures [50, 51]. In several applications, the vector of interest is a sparse vector (or has a sparse representation) or the matrix of interest is a low rank matrix. Recently, some interesting ideas have been introduced to leverage these low dimensional structures [52, 53]. For instance, it has been shown that sparse vectors can be recovered from a small number of non-adaptive measurements. It means that, if we want to recover an Ndimensional sparse vector, we do not need a set of N independent linear measurements [52, 54]. In [53], this paper presents a new randomized approach to high-dimensional low rank (LR) plus sparse matrix decomposition. For a data matrix $\mathbf{D} \in \mathbb{R}^{N_1 \times N_2}$, the complexity of conventional decomposition methods is $\mathcal{O}(N_1 N_2 r)$, which limits their usefulness in big data settings (r is the rank of the LR component). In addition, the existing randomized approaches rely for the most part on uniform random sampling, which may be inefficient for many real-world data matrices. The proposed subspace learning-based approach recovers the LR component using only a small subset of the columns/rows of data and reduces complexity to $\mathcal{O}(\max(N_1, N_2) r^2)$. Even when the columns/rows are sampled uniformly at random, the sufficient number of sampled columns/rows is shown to be roughly $\mathcal{O}(r\mu)$, where μ is the coherency parameter of the LR component. In addition, efficient sampling algorithms are proposed to address the problem of column/row sampling from structured data. Moreover, this paper presented a scalable iterative algorithm for column/row sampling form in a sparse manner. It can be used for corrupted data matrices that their computation complexity and memory requirement are roughly linear with $r \max(N_1, N_2) r$. In [51], in subspace clustering, a group of data points belonging to a union of subspaces are assigned membership to their respective subspaces. This paper presents a new approach dubbed Innovation Pursuit (iPursuit) to the problem of subspace clustering using a new geometrical idea whereby each subspace is identified based on its novelty with respect to the other subspaces. The proposed approach finds the subspaces consecutively by solving a series of simple linear optimization problems, each searching for some direction in the span of the data that is potentially orthogonal to all subspaces except for the one to be identified in one step of the algorithm. A detailed mathematical analysis is provided establishing sufficient conditions for the proposed approach to correctly cluster the data points. Remarkably, the proposed approach can provably yield exact clustering even when the subspaces have significant intersections under mild conditions on the distribution of the data points in the subspaces. Moreover, it is shown that the complexity of iPursuit is almost independent of the dimension of the data. The numerical simulations demonstrate that iPursuit can often outperform the state-of-the-art subspace clustering algorithms, more so for subspaces with significant intersections.

The paper [52] derived an interesting result which shows that a sparse vector can be recovered from a small set of random orthogonal projection. Suppose that $\mathbf{s} \in \mathbb{R}^N$ is a sparse vector and the number of its non-zero elements is equal to $\|\mathbf{s}\|_0$. Define $\mathbf{T} \in \mathbb{R}^{N \times N}$ as a matrix which contains a set of orthonormal basis. It means

$$\mathbf{T}^T \mathbf{T} = \mathbf{I}, \qquad (3.30)$$

where $\mathbf{I}$ is the identity matrix. Now suppose that $\mathbf{U} \in \mathbb{R}^{N \times m}$ is set of randomly chosen columns of $\mathbf{T}$. In addition, define μ as the minimum value which satisfies

$$\max_i \|\mathbf{U}^T \mathbf{e}_i\|_2 \leq \frac{\mu m}{N}, \qquad (3.31)$$

If the parameter μ is small, it means that the columns subspace of the matrix $\mathbf{U}$ is not aligned with the standard basis. The orthogonal matrix $\mathbf{U}$ is used to measure the sparse vector. Thus, it is essential that $\mathbf{U}$ not to be a sparse matrix. In [52], it was shown that if

$$m \geq c\|\mathbf{s}\|_0 \mu \log \frac{N}{\delta} \qquad (3.32)$$

then the optimal point of

$$\min_{\hat{\mathbf{z}}} \quad \|\hat{\mathbf{z}}\|_1$$
$$\text{subject to} \quad \mathbf{U}^T \hat{\mathbf{z}} = \mathbf{U}^T \mathbf{s}. \qquad (3.33)$$

is equal to $\mathbf{s}$ with probability at least $(1 - \delta)$, where c is a constant number. It means that an N-dimensional sparse vector can be recovered from a small random non-adaptive linear measurements and the sufficient number of measurements is linearly dependent on the number of non-zero elements of the sparse vector (or if it is not exactly sparse, the number of dominant elements [55]). As it has been mentioned, in order to model the noise vector, we update the DC power flow model to

$$\mathbf{p} = \mathbf{B}\delta + \boldsymbol{\epsilon}, \qquad (3.34)$$

where $\boldsymbol{\epsilon}$ is the so-called noise vector. In this chapter, it is assumed that the additive noise vector is sparse. We exploit this geometrical structure to ameliorate the quality of the Sparse Vector Recovery algorithm.

3.5.2 Proposed Sparsity-Based DC Power Flow Estimation

Due to the existence of a slack bus, we have at least a row that is linearly dependent on the other rows of the $\mathbf{B}$ matrix; hence, the matrix $\mathbf{B}$ is not a full rank matrix in most configurations. Suppose that the rank of the matrix $\mathbf{B}$ is equal to r_B. Then, (3.34) can be rewritten as

$$\mathbf{p} = \mathbf{Q}\mathbf{a} + \boldsymbol{\epsilon}, \qquad (3.35)$$

where $\mathbf{Q} \in \mathbb{R}^{\aleph \times r_B}$ is an orthonormal basis for the columns subspace of $\mathbf{B}$. The conventional method for estimating the coefficient vector $\mathbf{a}$ is the least-squares method, which solves the following optimization problem

$$\min_{\hat{\mathbf{a}}} \|\mathbf{p} - \mathbf{Q}\hat{\mathbf{a}}\|_2. \tag{3.36}$$

The least-squares approach to (3.36) projects the vector $\mathbf{p}$ onto the columns subspace of $\mathbf{Q}$. Therefore, the quality of the least-square estimator is function of the projection of the noise vector ϵ on the columns subspace of $\mathbf{Q}$. As a result, the least squares approach does not yield accurate estimation if the noise vector has a large projection on the columns subspace of $\mathbf{Q}$.

Here, we consider cases with the error vector ϵ being relatively sparse. l_1-minimization algorithms are robust to additive sparse error vector with arbitrarily large magnitudes [55, 56]. Specifically, if ϵ is sufficiently sparse and $\mathbf{Q}$ meets the incoherence conditions [50, 56], then the optimal point of

$$\min_{\hat{\mathbf{a}}} \|\mathbf{p} - \mathbf{Q}\hat{\mathbf{a}}\|_1 \tag{3.37}$$

is the same as $\mathbf{a}$, and the optimal point is independent from the magnitude of ϵ.

In the proposed method, we assume that $\mathbf{B}$ is not a full rank matrix. Hence, $r_B < \aleph$. We define $\mathbf{Q}^{\perp} \in \mathbb{R}^{\aleph \times (\aleph - r_B)}$ as the matrix whose columns subspace is the complement of the columns subspace of $\mathbf{Q}$. It has been demonstrated in [50, 56] that (3.37) is equivalent to

$$\min_{\hat{\epsilon}} \quad \|\hat{\epsilon}\|_1$$
$$\text{subject to} \quad (\mathbf{Q}^{\perp})^T \hat{\epsilon} = (\mathbf{Q}^{\perp})^T \mathbf{p}, \tag{3.38}$$

to mention that, if $\mathbf{a}_o$ is the optimal point of (3.37) and $\hat{\epsilon}_o$ is the optimal point of (3.38), then $\mathbf{p} - \mathbf{Q}\mathbf{a}_o = \hat{\epsilon}_o$. Consequently, according to (3.32), if

$$(\aleph - r_B) \geq c\|\epsilon\|_0 \mu_B \log \frac{N}{\delta} \tag{3.39}$$

then the optimal point of (3.38) is equal to $\mathbf{a}$ with the minimum probability of $(1-\delta)$, where $\|\epsilon\|_0$ is the number of non-zero elements of ϵ, and similar to (3.31), μ_B is described as

$$\max_i \|(\mathbf{Q}^{\perp})^T \mathbf{e}_i\|_2 \leq \frac{\mu_B(\aleph - r_B)}{\aleph}. \tag{3.40}$$

Therefore, if the rank of $\mathbf{B}$ is sufficiently small (i.e., $(\aleph - r_B)$ is sufficiently large), the l_1-minimization algorithm (3.37) can lead to exact estimation.

Henceforth, we utilize the aforementioned approach to find the noise vector ϵ in (3.34). Our approach is proposed for DC power flow calculation in transmission networks with measurements that are accurate for the most part, i.e., the noise vector is sparse and the DC power flow assumptions are considered to obtain linear model. Consequently, Sparsity-based DC Power Flow refers to solving the DC power flow problem as a linear problem with noisy measurement inputs utilizing sparsity-based vector decomposition. It also finds the noisy measured data to increase the accuracy of the DC power flow solution.

3.5.3   Case Study and Discussion

We provide some simulation results to validate the performance of the introduced Sparsity-based DC Power Flow Estimation approach. In [42], it has been shown that the proposed ℓ_1-norm minimization algorithm yields robustness against the large values of the sparse errors. The algorithm is applied to IEEE 118-bus and 300-bus test case systems [57]. In [42], it has been demonstrated that the performance of the algorithm is function of the number of dominant elements (or non-zero elements). Also, it has been shown that the proposed method outperforms the least-squares approach in terms of estimation accuracy.

Simulation Results

Here, we study the performance of the proposed algorithm under different sparsity levels. Suppose that the additive error vector is a sparse vector. Each element of the error vector is non-zero with probability equal to α. Thus, we control the sparsity level by adjusting the parameter α. Figures 3.2 and 3.3 show the estimated error vector for two different values of α for IEEE 300-bus test system [57].

It can be seen that the performance of the proposed algorithm improves when the number of non-zero elements of the error vector is decreased. This agrees with the analysis, which shows that the proposed estimator is only sensitive to the number of dominant elements (or non-zero elements) of the error vector. However the proposed algorithm is more accurate than the least-square approach, some of the errors can not be detected. For instance, as it has been shown in Fig. 3.4, there are two types of mistakes in the error detection:

- An error occurred but it has not been detected by the proposed approach;
- Our approach has detected an error at a specific bus. However, there is no error at that bus.

In this section, we provided more results to describe the proposed approach thoroughly. According to the simulation results provided here and in [42], the performance of this method is not affected by the magnitude of the measurement errors. In [42], we have evaluated the accuracy of the proposed error detection approach by defining six scenarios based on IEEE 118-bus and IEEE 300-bus test systems. In this scenario we designedly add some sparse measurement noise to the

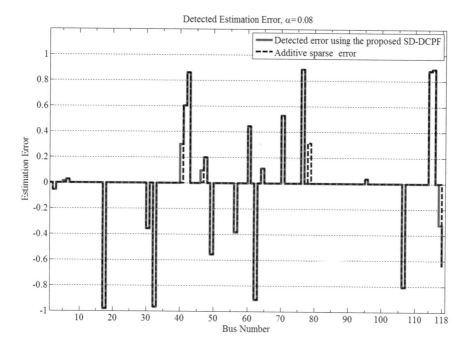

Fig. 3.2 Detected estimation error, IEEE 300-bus test network, $\alpha = 0.08$

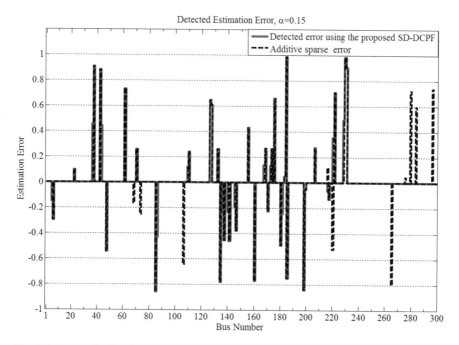

Fig. 3.3 Detected estimation error, IEEE 300-bus test network, $\alpha = 0.15$

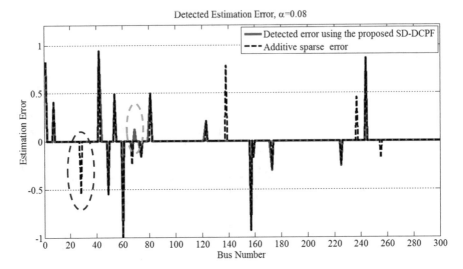

Fig. 3.4 Detected estimation error and erroneous detected error, IEEE 118-bus test network, $\alpha = 0.08$

measured data vector. The probability that an element of the sparse error vector is non-zero set to be 3, 8, and 15%. By reducing the sparsity of the noise vector, i.e. increasing the number of non-zero elements, the detection rate will decrease. In other words, lower sparsity in the noise vector leads to lower accuracy in detecting the noisy data. In order to guarantee that the proposed decomposition method obtains the exact sparse vector, the column subspace of **B** matrix should not be aligned with the standard basis.

References

1. V.C. Gungor et al., Smart grid technologies: communication technologies and standards. IEEE Trans. Ind. Inf. **7**(4), 529–539 (2011)
2. W. Su, H.R. Eichi, W. Zeng, M.-Y. Chow, A survey on the electrification of transportation in a smart grid environment. IEEE Trans. Ind. Inf. **8**(1), 1–10 (2012)
3. M.H. Amini, A. Islam Allocation of electric vehicles' parking lots in distribution network, , in *Proceedings of IEEE Innovative Smart Grid Technologies Conference (ISGT)*, Washington, DC, Feb 2014, pp. 1–5
4. M.H. Amini, O. Karabasoglu, M.D. Ilić, K.G. Boroojeni, S.S. Iyengar, ARIMA-based demand forecasting method considering probabilistic model of electric vehicles' parking lots. IEEE PES General Meeting 2015, Denver, CO, 26–30 July 2015
5. M.H. Amini, A.I. Sarwat, Optimal reliability-based placement of plug-in electric vehicles in smart distribution network. Int. J. Eng. Sci. **4**(2), 43–49 (2014)
6. J.M. Carrasco et al., Power-electronic systems for the grid integration of renewable energy sources: a survey. IEEE Trans. Ind. Electron. **53**(4), 1002–1016 (2006)

7. X. Yu, C. Cecati, T. Dilon, M.G. Simoes, The new frontier of smart grids. IEEE Ind. Electron. Mag. **5**(3), 49–63 (2011)
8. M.H. Amini, J. Frye, Marija D. Ilić, O. Karabasoglu, Smart residential energy scheduling utilizing two stage mixed integer linear programming, in *IEEE 47th North American Power Symposium (NAPS 2015)*, Charlotte, NC, 4–6 Oct 2015
9. National Institute of Standards and Technology, NIST framework and roadmap for smart grid interoperability standards, release 1.0. Office of the National Coordinator for Smart Grid Interoperability-U.S. Department of Commerce, NIST Special Publication 1108, Jan 2010
10. S. Kar, G. Hug, J. Mohammadi, J.M.F. Moura, Distributed state estimation and energy management in smart grids: a consensus+innovations approach. IEEE J. Sel. Top. Sign. Proces. **99**, 1–16 (2014)
11. F. Kamyab, M.H. Amini, S. Sheykhha, M. Hasanpour, M.M. Jalali, Demand response program in smart grid using supply function bidding mechanism. IEEE Trans. Smart Grid **7**(3), 1277–1284 (2016)
12. M.H. Amini, M.P. Moghaddam, Probabilistic modelling of electric vehicles' parking lots charging demand, in *21th Iranian Conference on Electrical Engineering ICEE2013*, Ferdowsi University of Mashhad, 14–16 May 2013
13. A. Zidan, E.F. El-Saadany, A cooperative multi-agent framework for self-healing mechanisms in distribution systems. IEEE Trans. Smart Grid **3**(3), 1525–1539 (2012)
14. R.E. Brown, Impact of smart grid on distribution system design , in *Proceedigs IEEE Power and Energy Society General Meeting*, Pittsburgh, PA, July 2008, pp. 1–4
15. M.H. Amini, B. Nabi, M.-R. Haghifam, Load management using multi-agent systems in smart distribution network, in *Proceedings of IEEE Power and Energy Society General Meeting*, Vancouver, BC, July 2013, pp. 1–5
16. S. Bera, S. Misra, P.C. Rodriguez, Cloud computing applications for smart grid: a survey . IEEE Trans. Parallel Distrib. Syst. **99**, 1–18 (2014)
17. C.-T. Yang, W.-S. Chen, K.-L. Huang, J.-C. Liu, W.-H. Hsu, C.-H. Hsu, Implementation of smart power management and service system on cloud computing , in *Proceedings of IEEE International Conference on UIC/ATC*, 2012, pp. 924–929
18. M. Kezunovic, X. Le, G. Santiago, The role of big data in improving power system operation and protection, in *IEEE Bulk Power System Dynamics and Control-IX Optimization, Security and Control of the Emerging Power Grid (IREP), 2013 IREP Symposium*, 2013
19. Y. Shoham, K. Leyton-Brown, *Multi-Agent Systems: Algorithmic*. Game Theoretic and Logical Foundations. (Cambridge University Press, Cambridge, 2009–2010)
20. F. Bellifemine, G. Caire, D. Greenwood, *Developing Multi-agent systems with JADE* (Wiley, New York, 2007)
21. M. Wooldridge, G. Weiss, Intelligent agents, in *Multi-Agent Systems* (MIT Press, Cambridge, MA, 1999), pp. 3–51
22. P. Siano, C. Cecati, H. Yu, J. Kolbusz, Real time operation of smart grids via FCN networks and optimal power flow. IEEE Trans. Ind. Inf., **8**(4), 944–952 (2012)
23. A.I. Sarwat, M.H. Amini, A. Domijan Jr., A. Damnjanovic, F. Kaleem, Weather-based interruption prediction in the smart grid utilizing chronological data. J. Mod. Power Syst. Clean Energy **4**(2), 308–315 (2016)
24. J.D. Glover, M.S. Sarma, *Power System Analysis and Design*, 3rd edn. (Pacific Grove, CA, Brooks/Cole, 2002)
25. B. Stott, Review of load-flow calculation methods . Proc. IEEE **62**, 916–929 (1974)
26. A.J. Wood, B.F. Wollenberg, *Power Generation, Operation and Control*, 2nd edn. (Wiley, New York, 1996)
27. G. Giannakis, V. Kekatos, N. Gatsis, S.-J. Kim, H. Zhu, B. Wollenberg, Monitoring and optimization for power grids: a signal processing perspective. IEEE Signal Process. Mag. **30**(5), 107–128 (2013)
28. L. Powell, DC load flow, Chap. 11, in *Power System Load Flow Analysis*. McGrawHill Professional Series (McGrawHill, New York, 2004)

29. B. Stott, J. Jardim, O. Alsac, DC power flow revisited. IEEE Trans. Power Syst. **24**(3), 1290–1300 (2009)
30. R.J. Kane, F.F. Wu, Flow approximations for steady-state security assessment. IEEE Trans. Circuits Syst. **CAS-31**(7), 623–636 (1984)
31. R. Baldick, Variation of distribution factors with loading . IEEE Trans. Power Syst. **18**(4), 1316–1323 (2003)
32. L. Xiong, W. Peng, L. Pohchiang, A hybrid AC/DC microgrid and its coordination control. IEEE Trans. Smart Grid **2**(2), 278–286 (2011)
33. M.D. Ilić, J. Zaborszky, *Dynamics and Control of Large Electric Power Systems* (Wiley, New York, 2000)
34. S.M. Kay, *Fundamentals of Statistical Signal Processing: Estimation Theory*, 1st edn. (Prentice-Hall International Editions, Englewood Cliffs, 1993)
35. F.F. Wu, K. Moslehi, A. Bose, Power system control centers: past, present, and future. Proc. IEEE **93**(11), 1890–1908 (2005)
36. F.C. Schweppe, J. Wildes, D.B. Rom, Power system static state estimation, Parts I, II and III. IEEE Trans. Power Apparatus Syst. **PAS-89**(1), 120–135 (1970)
37. A. Abur, A.G. Exposito, *Power System State Estimation: Theory and Implementation* (CRC Press, New York, 2002)
38. P.A. Ruiz, P.W. Sauer, Voltage and reactive power estimation for contingency analysis using sensitivities. IEEE Trans. Power Syst. **22**(2), 639–647 (2007)
39. K.G. Boroojeni, S. Mokhtari, M.H. Amini, S.S. Iyengar, Optimal two-tier forecasting power generation model in smart grid. Int. J. Inf. Process. **8**(4), 1–10 (2014)
40. M.H. Amini, A.I. Sarwat, S.S. Iyengar, I. Guvenc, Determination of the minimum-variance unbiased estimator for dc power-flow estimation, in *40th IEEE Industrial Electronics Conference (IECON 2014)*, Dallas, TX, 2014
41. M.H. Amini, M.D. Ilić, O. Karabasoglu, DC power flow estimation utilizing Bayesian-based LMMSE Estimator, in *IEEE PES General Meeting 2015*, Denver, CO, 26–30 July 2015
42. M.H. Amini et al., Sparsity-based error detection in DC power flow state estimation. arXiv preprint arXiv:1605.04380, 2016
43. R.L. Rabiner, R.W. Schafer, *Digital Processing of Speech Signals* (Prentice Hall, Englewood Cliffs, 1978)
44. L.R. Bahl et al., Maximum mutual information estimation of hidden Markov model parameters for speech recognition, in *Proceedings of IEEE International Conference on Acoustics, Speech, and Signal Processing, ICASSP*, 1986
45. V. Digalakis, J.R. Rohlicek, M. Ostendorf. ML estimation of a stochastic linear system with the EM algorithm and its application to speech recognition. IEEE Trans. Speech Audio Process. **1**(4), 431–442 (1993)
46. V. Tarokh et al., Space-time codes for high data rate wireless communication: performance criteria in the presence of channel estimation errors, mobility, and multiple paths. IEEE Trans. Commun. **47**(2), 199–207 (1999)
47. O. Edfors et al., OFDM channel estimation by singular value decomposition. IEEE Trans. Commun. **46**(7), 931–939 (1988)
48. Y. Li, L.J. Cimini Jr., N.R. Sollenberger, Robust channel estimation for OFDM systems with rapid dispersive fading channels. IEEE Trans. Commun. **46**(7), 902–915 (1998)
49. R. Togneri, Estimation theory for engineers, 30 Aug 2005. Online Available: http://staffhome. ecm.uwa.edu.au/00014742/teach/Estimation_Theory.pdf
50. M. Rahmani, G. Atia, A subspace learning approach to high dimensional matrix decomposition with efficient information sampling. arXiv preprint arXiv:1502.00182, 2016
51. M. Rahmani, G. Atia, Innovation pursuit: a new approach to subspace clustering. arXiv preprint arXiv:1512.00907, 2015
52. E. Candes, J. Romberg, Sparsity and incoherence in compressive sampling. Inverse Prob. **23**(3), 969 (2007)
53. M. Rahmani, G. Atia, High dimensional low rank plus sparse matrix decomposition. arXiv preprint arXiv:1502.00182, 2015

54. E.J. Candes, J. Romberg, T. Tao, Robust uncertainty principles: exact signal reconstruction from highly incomplete frequency information. IEEE Trans. Inf. Theory **52**(2), 489–509 (2006)
55. E.J. Candes, T. Tao, Near-optimal signal recovery from random projections: universal encoding strategies? IEEE Trans. Inf. Theory **52**(12), 5406–5425 (2006)
56. E.J. Candes, T. Tao, Decoding by linear programming. IEEE Trans. Inf. Theory **51**(12) 4203–4215 (2005)
57. University of Washington Electrical Engineering, Power systems test case archive (2015). Online Available: http://www.ee.washington.edu/research/pstca/

Chapter 4
Bad Data Detection

One of the major roles that Smart Grid has promised to play is to provide a power to satisfy power demand with environmentally-friendly source of energy while maintaining an acceptable level of adequacy and security that traditional systems promise. As a result, there have been many efforts to develop estimation algorithms of the power system states which are the core of the time-sensitive grid management. In this chapter, we address auto-regressive load forecasting methods which play pivotal role in creating an accurate state estimator for the power grid management.

4.1 Preliminaries on Falsification Detection Algorithms

A time series is a sequence of data points, typically consisting of successive measurements made over a time interval. Time series are very frequently plotted via line charts and are used in any field which involves temporal measurements. Time series analysis comprises methods for analyzing time series data in order to extract meaningful statistics and other characteristics of the data. Time series forecasting is the use of a model to predict future values based on previously observed values. While regression analysis is often employed in such a way as to test theories that the current values of one or more independent time series affect the current value of another time series, autoregressive analysis is utilized to model the behavior of a time series without looking at any other time series [1].

© Springer International Publishing Switzerland 2017
K.G. Boroojeni et al., *Smart Grids: Security and Privacy Issues*,
DOI 10.1007/978-3-319-45050-6_4

4.1.1   Autocorrelation Function (ACF)

The ACF of a wide-sense stationary time series x_t is defined as function $AC(l) = \mathrm{corr}_l(x_t)$ where x_t and corr_l function is defined in the following way (μ and σ^2 are the mean and variance of x_t

$$\mathrm{corr}_l(x_t) = \mathbf{E}\left[\frac{(x_t - \mu)(x_{t+l} - \mu)}{\sigma^2}\right] \tag{4.1}$$

Moreover, the partial autocorrelation function (PACF) of a wide-sense stationary time series x_t is represented by $PAC(l)$ and is defined as follows:

$$PAC(l) = \begin{cases} AC(1), & l = 1; \\ \mathrm{corr}\left(x_{t+l} - P_{t,l}(x_{t+l}), x_t - P_{t,l}(x_t)\right), & l \geq 2. \end{cases} \tag{4.2}$$

where $P_{t,l}(x)$ denotes the projection of x onto the space spanned by $x_{t+1}, \ldots, x_{t+l-1}$.

Electricity load demand over a period of time is a seasonal non-stationary time series. Many attempts have been made by statisticians in order to model this kind of time series. As a result, in 2013, a step-by-step procedure has been developed [2] for applied seasonal non-stationary time series modeling following the Box–Jenkins Methodology which is a known modeling methodology first created by Box and Jenkins in 1976 [3] (Fig. 4.1).

4.2   Time Series Modeling of Load Power

The chronologically ordered power load values of a specific electrical system is a time series that occurs over a period of time in a seasonal manner. To model such time series as a function of its past values, analysts identify the pattern with the assumption that the pattern will persist in the future. Applying the Box–Jenkins methodology, this section emphasizes how to identify an appropriate time series model by matching behaviors of the sample ACF and PACF to the theoretical autocorrelation functions. In addition to model identification, the section examines the significance of the parameter estimates, checks the diagnostics, and validates the forecasts (Fig. 4.2).

4.2.1   Outline of the Proposed Methodology

We first need to check if the load data over time shows constant seasonal variance. To stabilize the variance, we apply an appropriate pre-differencing transformation (i.e., log, square root, etc.) on the original time series values. Next we examine

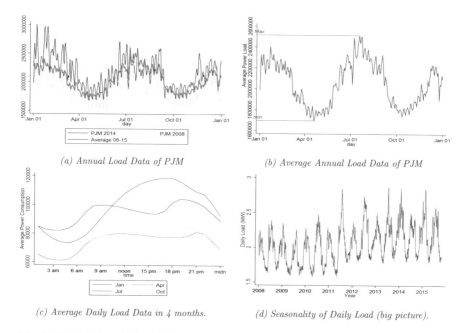

(a) Annual Load Data of PJM   (b) Average Annual Load Data of PJM

(c) Average Daily Load Data in 4 months.   (d) Seasonality of Daily Load (big picture).

Fig. 4.1 Statistical analysis of the load data of PJM network over a period of 8 years. (**a**) Annual load data of PJM. (**b**) Average annual load data of PJM. (**c**) Average daily load data in 4 months. (**d**) Seasonality of daily load (big picture)

the ACF of the transformed data both at non-seasonal and seasonal levels for any indication of stationary. Then, the time series values must be stationary where its mean and variance are constant through time. The constant mean and variance can be achieved by removing the pattern caused by the time dependent autocorrelation. Besides looking at the plot of the time series values over time to determine stationary or non-stationary, the sample ACF also gives visibility to the data. If the ACF of the time series values either cuts off or dies down fairly quickly, then the time series values should be considered stationary. On the other hand, if the ACF of the time series values either cuts off or dies down extremely slowly, then it should be considered non-stationary. The seasonal non-stationary time series values is turned into stationary time series values utilizing multiple differencing transformation. The transformed stationary timeseries is modeled to a moving average and/or autoregressive model based on how ACF and PACF of the time series behave. Finally, the detailed parameters of the model are estimated by trying multiple values and validating the forecasts using the Aikaike/Bayesian Information Criteria (AIC/BIC).

Fig. 4.2 The outline of the
methodology [1]

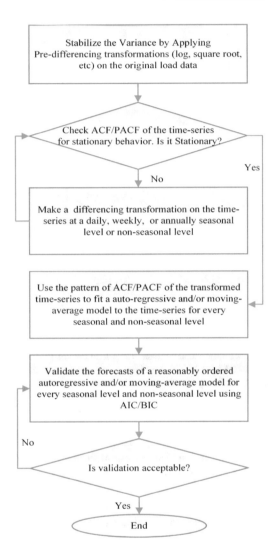

4.2.2   Seasonality

Any time series which shows periodic similarities over time is called to be
seasonal. There are different methods to deal with the seasonality of time series
and model it. A common seasonal model is Seasonal ARIMA model denoted by
ARIMA$(p, d, q)(P, D, Q)_m$, where m refers to the number of periods in each season,
and the uppercase P, D, Q refer to the autoregressive, differencing, and moving
average terms for the seasonal part of the ARIMA model.

The electrical load demand values follow some periodic patterns. For example,
the load demand on Sundays is similar to each other while they are different from

Mondays' load demand pattern; or, new year's holidays have a special load pattern every year. In order to deal with seasonality of time series, we need to check if the data over time shows constant seasonal variance. The seasonal behaviors appear at the exact seasonal lags L, 2L, 3L, 4L, and 5L. For daily data (L = 7), the exact seasonal lags are 7, 14, 21, and 28. For hourly data (L = 24), the exact seasonal lags would be 24, 48, 72, and 96. In general, the transformed time series values are considered stationary if the ACF shows both of the following behaviors [1]:

1. Cuts off or dies down fairly quickly at the non-seasonal level.
2. Cuts off or dies down fairly quickly at the seasonal level (exact seasonal lags or near seasonal lags).

Otherwise, these values are considered non-stationary.

Given that the time series values are considered stationary and exhibit behaviors (spikes) described in Table 4.1 Non- Seasonal Theoretical Box–Jenkins Models at the non-seasonal and seasonal levels, here is the three-step procedure for tentatively identifying a model:

STEP 1: Look at the behaviors (spikes) of the ACF and PACF at the non-seasonal level to identify a non- seasonal model.

STEP 2: Look at the behaviors (spikes) of the ACF and PACF at the seasonal level to identify a seasonal model.

STEP 3: Combine models from STEPs 1 and 2 to identify an overall tentatively model.

Once we obtain the overall tentatively model, we need to determine whether or not to include a constant term in a Box–Jenkins model. A constant term is the mean of the stationary time series values , which is equal (or nearly equal) or not equal to zero. In general, the rule of thumb is to include a constant mean in the model if the absolute value of $\zeta = \frac{\bar{z}}{s_z/\sqrt{n-b+1}}$ is greater than 2 where

$$\bar{z} = \frac{\sum_t z_t}{n-b+1}, s_z = \sqrt{\frac{\sum_t (z_t - \bar{z})^2}{n-b}} \tag{4.3}$$

Equivalently, if the p-value[1] associated with a constant mean is less than a significant level, we should include a constant mean in the model. Otherwise, we do not include it in the model.

In addition to the autocorrelation plots, the white noise output is provided to test the null hypothesis: the autocorrelations of the time series values are equal to zero. If it is rejected, the autocorrelations of the time series values are nonzero. In this case,

---

[1]p-value is a function of the observed sample results (a statistic) that is used for testing a statistical hypothesis. More specifically, the p-value is defined as the probability of obtaining a result equal to or "more extreme" than what was actually observed, assuming that the null hypothesis is true.

Table 4.1 Behavior of autocorrelation and partial-autocorrelation functions of both the hourly and daily load data in non-seasonal and multiple seasonal levels [1]

	Hourly load time series						Daily load time series				
	Non-seasonal (hourly)		Daily seasonality		Weekly seasonality		Non-seasonal (daily)		Weekly seasonality	
	ACF	PACF	ACF	PACF	ACF	PACF	ACF	PACF	ACF	PACF
Lag	Sine-wave dying down	Cut-off at Lag=1	Cut-off at Lag=2	Exp. dying down	Cut-off at Lag=1	Exp. dying down	Cut-off at Lag=2	Exp. dying down	Cut-off at Lag=1	Exp. dying down
1	0.735	0.735	−0.242	−0.334	−0.496	−0.659	−0.288	−0.33	−0.448	−0.637
2	0.474	−0.144	−0.155	−0.278	−0.012	−0.500	−0.154	−0.299	0.035	−0.458
3	0.263	−0.064	−0.041	−0.212	0.021	−0.389	−0.066	−0.254	0.009	−0.353
4	0.117	−0.028	0.003	−0.162	−0.030	−0.318	−0.032	−0.147	−0.077	−0.31
5	0.017	−0.033	0.057	−0.123	0.040	−0.239	0.099	−0.141	0.043	−0.257
6	−0.054	−0.048	0.131	−0.059	−0.031	−0.196	0.094	−0.068	0.007	−0.235

the white noise output shows that the null hypotheses for lags up to 6, 12, 18, and 24 are strongly rejected indicating that a time series model is needed. (For illustration, see Fig. 4.4 in Sect. 4.3.)

4.2.3 Fitting the AR and MA Models

Looking at the ACF and PACF of the stationary time series in Fig. 4.4e, we can identify an appropriate time series model by following the three-step procedure discussed in Seasonal Box–Jenkins Model Identification [1].

STEP 1: The ACF cuts off after lag 1 and the PACF dies down at the non-seasonal level indicate a first-order moving average model. Therefore the tentatively non-seasonal model is

$$z_t = \delta + a_t - \theta_1 a_{t-1} \tag{4.4}$$

STEP 2: The ACF cuts off after lag 12 and the PACF dies down at the seasonal level indicate a seasonal moving average model with lag 24. Therefore the tentatively seasonal model is

$$z_t = \delta + a_t - \theta_1 a_{t-24} \tag{4.5}$$

STEP 3: Combining models from STEP 1 and 2, the tentatively overall model is

$$z_t = \delta + a_t - \theta_{1,24} a_{t-24} \tag{4.6}$$

Note that if all the candidate models have the same k, then AICc and AIC will give identical (relative) valuations; hence, there will be no disadvantage in using AIC instead of AICc. Furthermore, if n is many times larger than k^2, then the correction will be negligible; hence, there will be negligible disadvantage in using AIC instead of AICc.

4.3 Case Study

In this section, our approach is implemented on the PJM network load data and the superior performance of our novel methodology is illustrated comparing the currently existing models. Throughout this section, we build both real-time and short-term forecaster for the PJM load data. The short-term forecaster is trained using the load data of 2008–2015, while the real-time model is trained using the load data of 2014 and 2015.

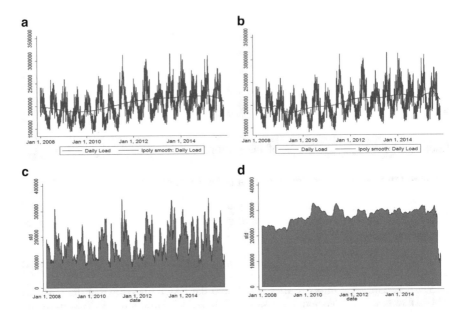

Fig. 4.3 Stabilizing the variance. (**a**) Best-fitted polynomial curve of degree 6 and the daily load values from 2008 to 2015. (**b**) Best-fitted polynomial curve of degree 16 and the daily load values from 2008 to 2015. (**c**) Monthly standard deviation of daily load values from 2008 to 2015. (**d**) Monthly standard deviation of stabilized daily load values

4.3.1   Stabilizing the Variance

As mentioned in the previous section, the first step for modeling the load data is to stabilize the load variance by transforming the original data. To do this, we first need to check whether the load data over time reflexes constant seasonal variance. Figure 4.3c shows that the daily load data has significant variance oscillation over the long-term horizon. To stabilize the variance, we subtract the best-fitted polynomial curve from the original data. As it has been shown in Fig. 4.3a, b, we can choose different polynomial curves with different degrees; however, the higher the degree is, the better-fit the curve will be. By subtracting the curve depicted in Fig. 4.3b from the original data, the data is transformed to a fairly constant variance signal (Fig. 4.3d).

In the next step, the time series values must be (wide-sense) stationary where its mean and auto-covariance do not vary with respect to time. One of the common ways to figure out if a time series is stationary is looking at its ACF curve. If the ACF of the original signal either cuts off or dies down fairly quickly, then the time series values should be considered stationary. On the other hand, if the ACF values either cuts off or dies down extremely slowly, then it should be considered non-stationary. Figure 4.4a shows the ACF of the hourly load data for the first 100 lags. The ACF function is a sine curve of period 24 which dies down very slowly. This means that the hourly load data is a seasonal time series of period 24 and of course, it is not a WSS time series. Almost any seasonal non-stationary original signal will be turned into stationary signal if multiple multi-lag differentiating[2] operators are applied to the signal in both seasonal and non-seasonal levels. In this case, we first try transformation $T_1 = 1 - L^{24}$ on the original hourly load data to remove the sine pattern of the ACF. As Fig. 4.4b shows, the ACF of T_1-transformed hourly load data dies down after about 75 lags. By applying transformation $T_2 = (1 - L^{168})(1 - L^{24})$ on the original data, we also remove another sine-pattern of the ACF because of weekly-seasonality of the original data which is illustrated in Table 4.2. As a result, the ACF of T_2-transformed hourly load data behaves like a quickly dying down curve (Fig. 4.4c). By applying the combination of a differentiating transformation and T_2 ($T_3 = (1 - L)T_2$) on the original data, the ACF curve looks more like the one of a WSS time series as it dies down quicker (Fig. 4.4d).

4.3.2 Fitting the Stationary Signal to a Model with Autoregressive and Moving-Average Elements

In this step, the transformed stationary time series is modeled to a moving average and/or autoregressive model based on the behavior of ACF and PACF of the time series. Table 4.1 shows the values of ACF and PACF functions in different non-seasonal and seasonal levels and for both transformed hourly and daily load time series.

4.3.2.1 Real-Time Forecasting Model

As you see in the left-hand side of this table which is dedicated to the transformed hourly load time series, the autocorrelation of the subsequent hours (the non-seasonal part of the time series) dies down in a Sine-wave manner with respect to the lag parameter (see the values of the second column of Table 4.1). Additionally, the partial autocorrelation of subsequent hours of the load-time series cuts off suddenly at lag $L = 1h$ (values of the third column of the same table supports the claim). The combinational behavior of the ACF and PACF functions at the non-seasonal level implies that the transformed hourly load values (η_t) follows an auto-regressive model of order one; i.e., the time series consists of the auto-regressive term $\phi_1^{(h)} \eta_{t-1}$ where $\phi_1^{(h)}$ denotes the coefficient of the auto-regressive term that will be computed in the model fine-tuning step (next step).

The values of the fourth and fifth columns of Table 4.1 illustrate that the autocorrelation of the transformed hourly load time series with daily lags ($L = 24, 48, 72, \ldots$) will substantially decrease after the first 2 days while its partial

---

[2]Operator D_k is called a k-lag differentiating operator if $D_k \cdot x_t = x_t - L^k \cdot x_t$.

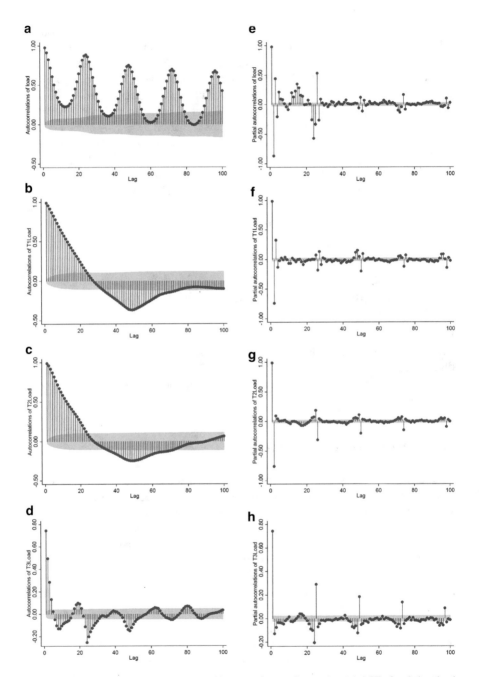

Fig. 4.4 Transforming the hourly load data into a stationary time series. (**a**) ACF of real-time load data. (**b**) ACF of T_1·Load. (**c**) ACF of T_2·Load. (**d**) ACF of T_3·Load. (**e**) PACF of real-time load data. (**f**) PACF of T_1·Load. (**g**) PACF of T_2·Load. (**h**) PACF of T_3·Load

Table 4.2 Comparing the day-ahead and real-time performance of the proposed forecaster [1]

Model	Training set	Forecasted	RMSE	MAPE
Real-time	2014	2015	0.67	0.49
Real-time	2014	2014–2015	0.63	0.43
Real-time	2014	2008–2015	0.66	0.47
Real-time	2014–2015	2014–2015	0.61	0.42
Real-time	2014	2008–2015	0.65	0.46
Day-ahead	2008–2014	2015	5.07	4.37
Day-ahead	2008–2014	2008–2015	4.98	4.32

autocorrelation dies down exponentially with respect to the daily lags. This behavior induces that the time series should include the moving-average of the second order in the daily seasonal level; i.e., the transformed hourly load data contains the summation of the following two moving-average signals which represent the non-deterministic identity of the time series: $\theta_1^{ma,d}\varepsilon_{t-24}^{(h)} + \theta_2^{(d)}\varepsilon_{t-48}^{h}$ where $\theta_1^{(d)}$ and $\theta_2^{(d)}$ represent the coefficients of the moving average terms of the model that will be specified in the next step of the model creation process. Moreover, $\varepsilon_t^{(h)}$ shows the non-deterministic part of the whole signal (η_t) and is traditionally called the "error value" of the ARMA model that will be built subsequently.

The values of the sixth and seventh columns of Table 4.1 reveal that the autocorrelation and partial autocorrelation of the transformed hourly load time series η_t with weekly lags ($L = 168, 336, \ldots$) show similar behavior as the ones with daily lags did. In fact, the values of ACF function will suddenly reduce after the first weekly lag while PACF dies down exponentially with respect to the weekly lags. This combination of behaviors leads us to the conclusion that the time series η_t should include the moving-average of the first order in the weekly seasonal level; i.e., the time series η_t has a non-deterministic moving-average part in the form of $\theta_1^{ma,w}\varepsilon_{t-168}^{(h)}$ where $\theta_1^{ma,w}$ represents the coefficient of the moving-average term with weekly lag and will be specified in the next step of the model creation process.

The result of Table 4.1 validates that the best real-time forecaster should have a moving-average element of lag one in the non-seasonal level, two auto-regressive lag at the daily seasonal level, and one auto-regressive lag at the weekly level:

$$\eta_t = \underbrace{\phi_1^{(h)}\eta_{t-1}}_{\text{non-seasonal}} + \underbrace{\theta_1^{(d)}\varepsilon_{t-24}^{(h)} + \theta_2^{(d)}\varepsilon_{t-48}^{(h)}}_{\text{daily-seasonal}} + \underbrace{\theta_1^{(w)}\varepsilon_{t-168}^{(h)}}_{\text{weekly-seasonal}} \qquad (4.7)$$

4.3.2.2 Day-Ahead Forecasting Model

As you see in the right-hand side of this table which is dedicated to the transformed daily load time series, the autocorrelation of the subsequent days (the non-seasonal part of the time series) drops down suddenly after the second lag (see the values of the eighth column of Table 4.1). Additionally, the partial autocorrelation of subsequent days of the load-time series dies down exponentially (values of the ninth column of the same table supports the claim). The combinational behavior of the

ACF and PACF functions at the non-seasonal level implies that the transformed daily load values (δ_t) follows a moving-average of order two; i.e., the time series consists of two moving-average terms $\gamma_1^{(d)}\varepsilon_{t-1}^{(d)} + \gamma_2^{(d)}\varepsilon_{t-2}^{(d)}$ where $\gamma_1^{(d)}$ and $\gamma_2^{(d)}$ denote the two coefficients of the auto-regressive terms that will be computed in the model fine-tuning step (next step). Moreover, $\epsilon_t^{(d)}$ shows the non-deterministic part of the whole signal (δ_t) and is traditionally called the "error value" of the ARMA model that will be built subsequently.

The values of the last two columns of Table 4.1 illustrate that the autocorrelation of the transformed daily load time series with weekly lags ($L = 7, 14, 21, \ldots$) will substantially decrease after the first week while its partial autocorrelation dies down exponentially with respect to the weekly lags. This behavior induces that the time series should include the moving-average of order one in the weekly seasonal level; i.e., the transformed daily load data contains the following moving-average signal which represents the non-deterministic identity of the time series: $\gamma_1^{(w)}\varepsilon_{t-7}^{(d)}$ where $\gamma_1^{(w)}$ represents the coefficient of the moving-average term of the model that will be specified in the next step of the model creation process.

The result of Table 4.1 validates that the best real-time forecaster should have a moving-average element of lag two in the non-seasonal level and one moving-average term at the weekly level [1]:

$$\delta_t = \underbrace{\gamma_1^{(d)}\varepsilon_{t-1}^{(d)} + \gamma_2^{(d)}\varepsilon_{t-2}^{(d)}}_{\text{non-seasonal}} + \underbrace{\gamma_1^{(w)}\varepsilon_{t-7}^{(d)}}_{\text{weekly-seasonal}} \qquad (4.8)$$

4.3.3   Model Fine-Tuning and Evaluation

Prior to this step, the orders of auto-regression (AR) and moving-average (MA) elements of the model have been set to their default values specified by the cut-off points of the ACF and PACF functions. Here, the model is fine-tuned by repeatedly changing the AR/MA orders of the model (by unity-step) and evaluating the model-fitness using the Aikaike/Bayesian Information Criteria (AIC/BIC). See Fig. 4.5. Table 4.3 summarizes the performance of our model comparing the other forecasters of current literature (Figs. 4.6, 4.7, 4.8, 4.9, 4.10).

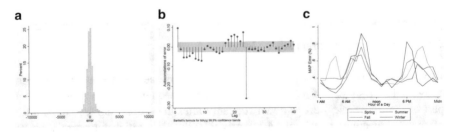

Fig. 4.5 Statistical analysis of hourly error. (**a**) Error histogram (real-time forecasting). (**b**) Correlogram of error for real-time forecasting. (**c**) Correlogram of error for real-time forecasting

Table 4.3 Comparison the accuracy of forecasters presented in Lauret et al. [2] and Chitsaz et al. [2]

Work	Year	Location	Method	White noise	MAPE	RMSE	MAE	Forecasting	AIC	BIC
Lauret et al.	2008	France	Bayesian method	No	3.17	N/A	N/A	Real-time	N/A	N/A
Lauret et al.	2008	France	Classical Method	No	10.28	N/A	N/A	Real-time	N/A	N/A
Chitsaz et al.	2015	British Columbia	WNN	No	1.81	2.46	N/A	Real-time	N/A	N/A
Chitsaz et al.	2015	California	WNN	No	2.57	3.67	N/A	Real-time	N/A	N/A
Chitsaz et al.	2015	British Columbia	SRWNN	No	1.69	2.29	N/A	Real-time	N/A	N/A
Chitsaz et al.	2015	California	SRWNN	No	2.37	3.38	N/A	Real-time	N/A	N/A
Boroojeni et al. [1]	2016	PJM	Box–Jenkins	Yes	3.65	4.77	3.53	Day-ahead	73742.2	73731.3
Boroojeni et al. [1]	2016	PJM	Box–Jenkins	Yes	1.11	1.45	1.06	Real-time	73851.2	73842.3

WNN: Wavelet Neural Network
SRWNN:Self-Recurrent WNN

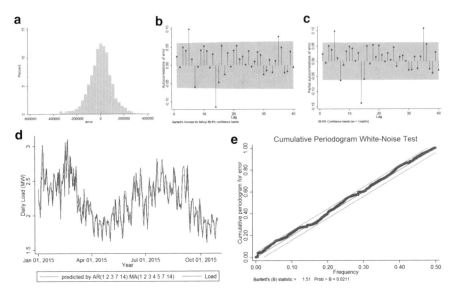

Fig. 4.6 Fitting an ARMA model to the transformed stationary signal and evaluating the best-fit model using AIC/BIC. (**a**) Forecasted plot using AR(1,2,3,7,14) MA(1,2,3,4,5,7,14). (**b**) Error correlogram of model AR(1,2,3,7,14) MA(1,2,3,4,5,7,14). (**c**) Error partial correlogram of model AR(1,2,3,7,14) MA(1,2,3,4,5,7,14). (**d**) Forecasted plot using AR(1,2,3,7,14) MA(1,2,3,4,5,7,14). (**e**) Periodogram of model AR(1,2,3,7,14) MA(1,2,3,4,5,7,14)

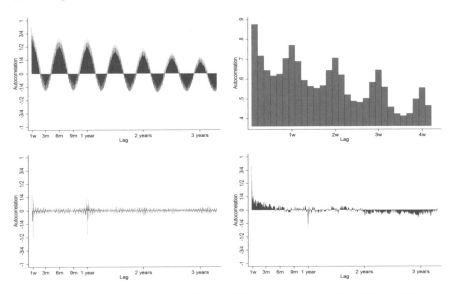

Fig. 4.7 Autocorrelation plots of the load data of PJM network over a period of 8 years [4]

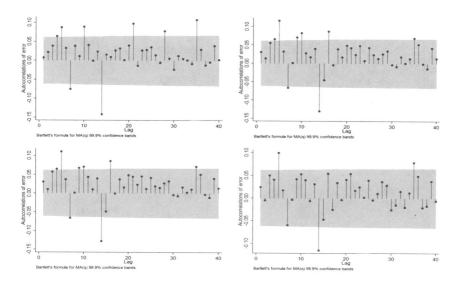

Fig. 4.8   Autocorrelation plots of the error values in four different cases

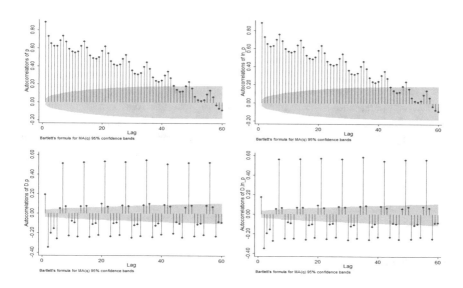

Fig. 4.9   Autocorrelation plots of the transformed load of PJM historical load values

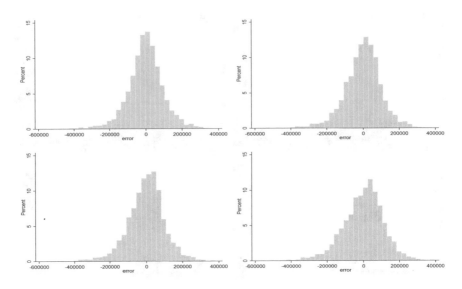

Fig. 4.10 Histograms of the error of forecasting models

4.4   Summary and Conclusion

Short-term load forecasting is essential for detecting falsification attacks in power systems. It is particularly more important for deregulated power systems where the forecast inaccuracies have significant implications for market operators, transmission owners, and market participants. An accurate load forecast results in establishing appropriate operational practices and bidding strategies as well as scheduling adequate energy transactions. This chapter presented a general technique of how to build an autoregressive model for the time series of load power demand in order to create an accurate short-term forecaster with the best Akaike/Bayesian Information Criteria (AIC/BIC). And the end, falsification detection algorithm using this technique was implemented for PJM/ERCOT real data.

References

1. K.G. Boroojeni, M.H. Amini, S. Bahrami, S.S. Iyengar, A.I. Sarwat, O. Karabasoglu, A novel multi-time-scale modeling for electric power demand forecasting: from short-term to medium-term horizon. Electr. Power Syst. Res. **142**, 58–73 (2017)
2. Theresa Hoang Diem NGO, Warner Bros. Entertainment Group, The Box–Jenkins methodology for time series models, SAS Global Forum, 2013
3. G.E.P. Box, G.M. Jenkins, *Time-Series Analysis: Forecasting and Control* (Holden-Day, CA, 1976)
4. W. Charytoniuk, M.-S. Chen, Very short-term load forecasting using artificial neural networks. IEEE Trans. Power Syst. **15**(1), 263–268 (2000)

Part II
Information Network Security

Chapter 5
Cloud Network Data Security

The recent advances in cloud computing have substantially changed people's understanding of computing hardware/software infrastructure and development methods. This fast transition to the clouds has coincided with the transition of mainframe computers to client/server models which has facilitated the utilization of cloud computing in different enterprises. This world wide transition has caused serious concerns regarding the confidentiality, integrity, and availability of communication and information systems participating in the cloud models as relocating data through the communication systems to the clouds causes various security and privacy issues that did not exist in traditional models before. This chapter proposes an Oblivious Routing-based Security Scheme ($\mathcal{ORSS}$) which effectively addresses the security issues caused by DDoS zombie attacks related to the data communication in cloud systems. The proposed security scheme is mathematically proved to guarantee some security low-threshold in long-term run. The simulation results on a general case study support the theoretical bound while showing that the proposed solution enhances the cloud system response time.

5.1 Introduction

Cloud computing has become a growing architectural model for organizations seeking to decrease their computer systems maintenance costs by migrating their computational tasks to third party organizations who offer software-as-a-service, platform-as-a-service, etc. As the result of such transition, many critical security concerns have emerged among the clients of clouds including data blocking and data leakage in the form of Distributed Denial of Service (DDoS) attacks [1]. As the cloud customers may have possibly thousands of shared resources in a cloud environment, the likelihood of DDoS attack is much more than a private architecture [2].

© Springer International Publishing Switzerland 2017
K.G. Boroojeni et al., *Smart Grids: Security and Privacy Issues*,
DOI 10.1007/978-3-319-45050-6_5

One of the main goals of DDoS attacks in cloud environments is to exhaust computer resources especially network bandwidth and CPU time so that the cloud service gets unavailable for the real legitimate clients [2]. In a general DDoS attack, the attacker usually mimics the legitimate web data traffic pattern to make it difficult for the victims to identify the attackers. This type of attack called "zombie attack" is a common type in cloud computing and is widely considered to be successful in circumventing the big portion of attack detection algorithms which work based on the abnormality of the traffic pattern generated by the DDoS attackers [2–4].

In recent years, many attempts have been made to mitigate the DDoS zombie attacks in general or specifically in the cloud environments. As we will see later in the literature review, these attempts have all relied on the conventional Internet routing algorithms and protocols (e.g., link-state, distance-vector) to route the data flow through the Internet. To the best of our knowledge, this is the first try ever made to mitigate the DDoS attacks by not utilizing the conventional Internet routing algorithms and creating a novel overlay network routing scheme based on the oblivious network design principles and algorithms. This project mainly focuses on the mitigation of the zombie attack and its destructive effect on the cloud services. The proposed mechanism is compatible with many of the DDoS detection and mitigation algorithms and can be integrated with them in order to enhance the security.

Related Work

Here, we review some of the recent studies on how to detect and mitigate the DDoS attacks in general and specifically in cloud environments.

Detection Mechanisms

The most common type of DDoS attacks in Internet is flooding. This popular attack has made many attempts in order to create effective countermeasures against it [5–7]. The designed/implemented countermeasures have an attack detection algorithm which predicts the occurrence of attacks utilizing the fact that the pattern of data traffic made by the hackers is deviated from the normal (natural) traffic pattern. The greater this deviation is, the higher the precision of the attack detection algorithms are. In 2015, Wang et al. proposed a graphical model in order to detect the security attacks (distributed denial of service) and inferring the probability distribution of attack [8].

As a response to this type of counter attacks, the attackers may regulate their generated data traffic pattern in a way that it no longer distinguishable with the other network data traffic and confuses the attack detection algorithms. Zombie attack is a good example in which the attackers perform DDoS attacks through a Poisson process in order to confuse the detection algorithms [3, 4]. Joshi et al. designed a cloud trace back model in dealing with DDoS attacks and addressed its performance using back propagation neural network and experimentally showed that the model is useful in tackling DDoS attacks [2].

Mitigation Mechanisms

In 2013, Mishra et al. proposed Multi-tenancy and Virtualization as two major solutions to mitigate the DDoS attacks in cloud environments [9]. Zissis et al. proposed a security solution to the security issues which helps the cloud clients make sure of their security by trusting a Third Party in a way that trust is created in a top-down manner where each layer of the cloud system trusts the layer lying immediately below it [10]. In 2011, Lua et al. reduced the possibility of DDoS attacks by utilizing a transparent and intelligent fast-flux swarm network which is a novel, efficient, and secure domain naming system.

This chapter proposes an Oblivious Routing-based Security Scheme ($\mathcal{ORSS}$) which mitigates the DDoS attack and related performance issues effectively while routing data through the Internet connecting the cloud servers and customers together. An oblivious routing algorithm usually proposes an overlay network in the form of a spanning tree or a set of trees on the graph representing the network [11]. These schemes are usually pretty versatile to the time-varying network topology and dynamic traffic pattern between different source–sink pairs by routing the traffic flows in a distributed way over the network and preventing the edge/node congestion. Moreover, the routing cost in such algorithms is usually proved to be asymptotically bounded to some coefficient of the optimal/minimum cost; this coefficient which is called the *competitiveness ratio* of the algorithm (or its corresponding routing scheme) has a value of greater than one [11, 12].

In this chapter, by defining the routing cost of data flow as a representation of security attack risk, we utilize an oblivious routing algorithm of competitiveness ratio $O(\log^2 |V|)$ mentioned in [11] to design a secure overlay network routing scheme. The proposed security scheme is mathematically proved to guarantee some security low-threshold in long-term run. As the security attacks cause reduction in performance, we will show how $\mathcal{ORSS}$ enhances the system performance. The simulation results will support the theoretical bound.

5.2 Data Security Protection in Cloud-Connected Smart Grids

Consider that in a cloud system, the cloud servers are located in the private cloud which has a connecting point (cloud gateway) to the Internet. The cloud customers located in the public cloud communicate with the cloud servers via the cloud gateway. A two-way communication between customers and the gateway is done through a network of routers connected with pre-established TCP connections.

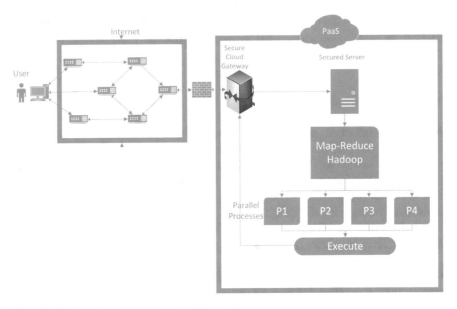

Fig. 5.1 The schematic representation of the problem scope

Let weighted graph $\mathcal{G} = (V, E, \lambda)$ denote the network connecting cloud secured gateway $g \in V$ and customers $c_1, c_2, \ldots, c_k \in V$ together where E specifies the set of TCP connections connecting the customers, gateway and the internal network nodes together and $\lambda : E \mapsto \mathbb{R}_{\geq 0}$ specifies the graph edge weights which is described later. Figure 5.1 depicts the cloud system, communication network, secured gateway, an example of cloud service (Platform as a Service) and the customer.

A security attack may be performed by a customer of a third party adversary by sniffing data or blocking data flow. Assuming that the data is encrypted by customer utilizing the elliptic curve cryptography and routed through the network of routers connected by TCP connections, it will make it very difficult for the hacker to identify the parameters used to draw the curve and then the point on the curve around which encryption is carried out. Furthermore, on receipt of the customer's task from the communicating network, the cloud gateway checks and verifies if any hacking took place. As a result, the security concern in such system is blocking data flow by the adversary in the communicating network (public cloud) before entering the cloud gateway through a distributed denial of service (DDoS) attack.

Attack Model

Distributed denial of service attacks are usually designed to be hard to detect. Attackers are mimicking network natural traffic statistical features to neutralize the detectors' effort which are based on these features. Zombie attack is a good example in which the attackers perform DDoS attacks through a Poisson process in order to confuse the detection algorithms [3, 4].

Consider random process $N_t^{(e)}$ denotes the number of security attacks occurred in TCP connection $e \in E$ during time interval $[0, t]$ for every $t > 0$. This is *homogeneous Poisson process* of intensity λ_e where λ is the weight function of the aforementioned weighted graph $\mathcal{G}$. We call λ_e as the *attack rate* in connection e. Note that for the sake of simplicity, we consider the attack rates as constant function of time; however, the proposed solution can be extended to the attacks with non-homogenous Poisson pattern.

Considering that there is a data flow of magnitude $f > 0$ is passing through the TCP connection e of transmission rate r_e, the data flow lasts for f/r_e time units. This implies that an average of $\lambda_e f/r_e$ takes place during the mentioned data flow session as the number of zombie attacks following a Poisson process. This value is formulated as the cost value incurred in edge e given a flow of f passing through it:

$$\text{cost}_e(f) = \frac{\lambda_e f}{r_e} \tag{5.1}$$

Assuming $\mathcal{N}_t$ as the random process of the total number of attack taking place in the TCP connections of the communicating network (i.e., $\mathcal{N}_t = \sum_e N_t^{(e)}$), then the expected value of the total number of attacks in the network, given that data flow of magnitude f_e is flowing through the connection e, in terms of cost value will then be equal to

$$\overline{\mathcal{N}}(f_\cdot) = \sum_{e \in E} \text{cost}_e(f_e) = \sum_{e \in E} \frac{\lambda_e f_e}{r_e} \tag{5.2}$$

Performance Evaluation

When an attack takes place in a TCP connection which flows customer c's data, c needs to send the data again through the network which increases the system response time to his query. Assuming that the customer considers path $p \subseteq E$ for the data of size f to flow toward the cloud secured gateway, the increased response time because of the attack is given by $\sum_{e \in p} f/r_e$. Assuming that query q has been sent by customer c through the path p, and the total response time of cloud servers to the query (excluding the communicating network delay) is ρ_q, the total response time is modeled by a random variable R_{pq} defined as follows (for the sake of simplicity and without loss of generality, assume that the query itself and its response are sent through the network in a data packet of unit size and the network transmission delay is the only considerable delay in the network):

$$R_{pq} = \rho_q + (2 + N_p) \sum_{e \in p} f_e/r_e \tag{5.3}$$

where N_p denotes the random variable counting the number of attacks taking place in path p given that flow f_e exists in edge e in the time that query q is submitted and responded by the cloud servers. Note that in Eq. (5.3), $2 + N_p$ specifies the

---

Algorithm 5.1: SCHEMECONSTRUCTOR [11]

---

1 for $i \leftarrow 1$ **to** $27 \log |V|$ **do**
2 | $\bar{H}^{(i)} \leftarrow$ HDS generated by FAKCHAROENPHOL;
3 | $T^{(i)} \leftarrow$ the HDT corresponding to $\bar{H}^{(i)}$;
4 end
5 for *source in G* **and** *sink in F* **do**
6 | **for** $i \leftarrow 1$ **to** $27 \log |V|$ **do**
7 | | **if** *source and sink are α-padded by* $\bar{H}^{(i)}$ **then**
8 | | | $T \leftarrow T^{(i)}$; $p \leftarrow$ the only path between the leaves source and sink in tree T;
        | | | **break**;
9 | | **end**
10 | **end**
11 | $\mathbb{S}(source, sink) \leftarrow$ the projection of p on graph $\mathcal{G}$;
12 end
13 return $\mathbb{S}$;

---

total number of times that query q is sent to the cloud server or responded back to the customer (data flow would exist exactly twice through path p if no attack occurs). This variable is the summation of $|p|$ Poisson distributed variables $N_p^{(e)}$ corresponding to the edges (e) participating in the path p such that $N_p^{(e)}$ has a mean of $\lambda_e \cdot f_e / r_e$. Henceforth, the expected value of the total system response time for arbitrary query q is obtained in the following way:

$$E[R_{pq}] = \rho_q + \left(2 + \sum_{e \in p} \frac{\lambda_e f_e}{r_e} \right) \sum_{e \in p} f_e / r_e \qquad (5.4)$$

This section proposes the novel $\mathcal{ORSS}$ method which utilizes a top-down oblivious routing scheme to solve security problem of the network connection between cloud server and its customers. Moreover, the method uses the optimization toolbox of MATLAB 2015 [13] made for minimizing the class of linearly constrained optimization problems. Here, we show how Algorithm 5.1 illustrates the construction of the oblivious routing scheme mentioned in [11].

Algorithm 5.1 gets the weighted graph $\mathcal{G}$ as its input and returns function $\mathbb{S} : \{G_i\}_1^n \times \{D_i\}_1^q \mapsto \mathcal{P}(E)$ which represents the oblivious routing scheme and specifies a path[1] in graph $\mathcal{G}$ for every pair of source–sink nodes. The algorithm starts with generating $27 \log |V|$ random HDSs utilizing randomized Algorithm 5.1. The reason for generating multiple random HDSs is to assure the existence of atleast one HDS $\bar{H}$ for every pair of nodes *source* $\in F$ and *sink* $\in G$. In fact, Iyengar et al.

---

[1]In this chapter, path of a graph is considered as a simple path and represented by a subset of edge set E such that there exist a permutation of edges in a path where the first edge is incident to the start node of the path, each two consecutive edges are incident to a common node, and the last edge is incident to the end node of the path.

in [11] proved that the probability of existing atleast one α-padding HDS among the $c \log |V|$ HDSs constructed by Algorithm 5.1 is more than $1 - e^{-\frac{(c-2)^2}{2c}}$ (for every $\alpha \leq 1/8$). By considering $c \geq 27$, the mentioned probability would be greater than $1 - 10^{-5}$ which provides a reasonable theoretical guarantee that Algorithm 5.1 will find atleast one HDS for every source–sink pair.

After creating $27 \log |V|$ random HDSs and their corresponding HDTs, Algorithm 5.1 runs its main loop in lines 5 to 11 for every pair of source–sink nodes. Assume T as the HDT corresponding to the α-padding HDS of an arbitrary pair of source–sink nodes. Tree T would have a pair of leaves (level-zero nodes) corresponding to the source and sink (as $\mathcal{G}$ is supposed to be a weighted graph of the weight function greater than one for every edge). Let p denote the only tree path connecting the mentioned leaves together. Finally, the resulted routing scheme is computed in line 11 where the path between source and sink nodes $\mathbb{S}(source, sink)$ is obtained by projecting p on graph $\mathcal{G}$. The projection of path p of HDT T on graph $\mathcal{G}$ is defined in the following way:

Consider γ_e as the shortest path between the incident nodes of arbitrary edge $e \in p$ in graph $\mathcal{G}$. The projection of tree path p on the graph is obtained by concatenating all of the shortest paths γ_es back to back. In the case that the concatenation result is not a simple path and has crossed some nodes more than once, the projected path will be the shortest simple path corresponding to the concatenation result.

As mentioned in Eq. (5.1), by considering the average number of attacks as the traffic cost of a given edge, function cost(e) is a sub-additive function of the data flow magnitude f_e. Consequently, regarding [11, 12], if data flow is routed by the aforementioned oblivious routing scheme $\mathbb{S}$, the total cost in all of the graph edges; i.e., the expected total number of attacks in the network connections doesn't exceed ψ times of the minimum possible expected attacks where $\psi = O(\log^2 |V|)$ (see Fig. 5.2 for more details).

In PaaS, the provider might give some control to the people to build applications on top of the platform. But any security below the application level such as host and network intrusion prevention will still be in the scope of the provider and the provider has to offer strong assurances that the data remains inaccessible between applications. PaaS is intended to enable developers to build their own applications on top of the platform. As a result it tends to be more extensible than SaaS, at the expense of customer-ready features. This trade-off extends to security features and capabilities, where the built-in capabilities are less complete, but there is more flexibility to layer on additional security.

Platform as a service (PaaS) is a category of cloud computing services that provides a platform allowing customers to develop, run, and manage Web applications without the complexity of building and maintaining the infrastructure typically associated with developing and launching an app. PaaS can be delivered in two ways: as a public cloud service from a provider, where the consumer controls software deployment and configuration settings, and the provider provides the networks, servers, storage, and other services to host the consumer's application; or as software installed in private data centers or public infrastructure as a service and managed by internal IT departments.

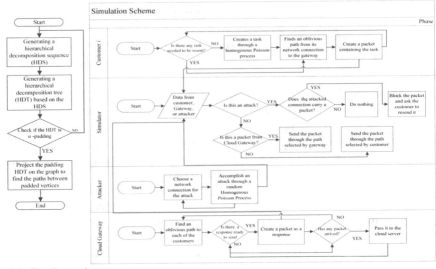

(a) Flowchart of
the oblivious rout-   (b) The Cross Flowchart of the simulation scheme.
ing algorithm

Fig. 5.2 The flowchart of oblivious routing algorithm. (**a**) Flowchart of the oblivious routing algorithm. (**b**) The cross flowchart of the simulation scheme

5.2.1 Simulation Scheme

Consider a cloud which provides its customers with a Platform as a Service (Paas). The customers have two-way communication with cloud's secured gateway through a network of routers connected by pre-established TCP connections (see Fig. 5.1). The network topology is represented by graph $G = (V, E)$ where V is the set of network routers and E is the TCP connections established between them. The secured gateway is connected to a server which is responsible to perform customers' tasks. The cloud secured server utilizes map-reduce model to perform customers tasks which are assumed to be oversized.

Assume that there are k customers spread over the network where the ith one sends the average of μ_i tasks to the cloud secured gateway within a homogeneous Poisson process. Each task is sent via a message of size $S_i \sim N(m_i, \sigma_i^2)$.

For every TCP connection $e \in E$, we consider r_e as its transmission rate and λ_e as the average number of security attacks which follows a homogeneous Poisson process.

Finally, we assume that every task takes $T \sim N(\tau, t^2)$ time to be performed and the result will be transferred back through network G as a message of size $S' \sim N(m', \sigma'^2)$.

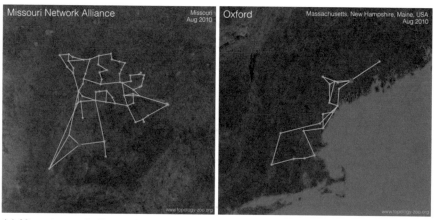

(a) Missouri network alliance. Location: Missouri, *(b) Oxford network. Location: MA, NH, MN, 2010*
2010 [22]. *[22].*

Fig. 5.3 Real-world network topologies used in the case study. The big vertex in each graph representation of network topologies specifies the location of cloud server. (**a**) Missouri network alliance. Location: Missouri, 2010 http://www.topology-zoo.org/dataset.html. (**b**) Oxford network. Location: MA, NH, MN, 2010 http://www.topology-zoo.org/dataset.html

5.2.2 Simulation Results

We implemented the proposed oblivious routing algorithm on two real-world network topologies obtained by Knight et al.'s study of network topology [14] shown in Fig. 5.3. The big red circular vertex represents the cloud gateway, while the other circles represent the cloud customers.

The following Table 5.1 specifies the performance of our novel routing scheme in contrast with the performance of the common traditional routing algorithms on the Oxford and Missouri network topologies. This table compares the percentage of successful attacks and the average increase in the system response time (obtained by Eq. (5.4)) because of DDoS zombie attacks. The comparison illustrates the superior performance of our novel scheme in both the average rate of successful attacks and response time increase because of zombie attacks. The algorithms that are compared to our novel oblivious routing algorithm are as follows:

- Safest path algorithm: the traditional single-source shortest path when the weight of each graph edge is considered as λ_e.
- Shortest path algorithm: the traditional single-source shortest path when the graph is assumed to be unweighted.

Finally, Fig. 5.4 compares the two aforementioned algorithms based on the probability distribution of success rate of zombie attacks (Fig. 5.4a–d) and the average expected network delay curve in the time horizon (Fig. 5.4e, f). As you see in the first two plots, our routing algorithm has a lower mode/median attack

Table 5.1 Comparing the percentage of successful attacks and the average increase in the system response time [obtained by Eq. (5.4)] because of DDoS zombie attacks

	Oblivious routing	Safest path routing	Shortest path routing
Oxford topology (n = 20)			
Average # of links in a path	6	5.75	4.8750
Average rate of successful attack (%)	7.2257	7.9765	9.2090
Response time increase because of attacks (%)	0.1036	0.1313	0.1611
Missouri topology (n = 67)			
Average # of links in a path	9.5417	6.9583	6.9167
Average rate of successful attack (%)	11.9940	12.1463	13.0464
Response time increase because of attacks (%)	0.3439	0.3525	0.3736

The comparison illustrates the superior performance of our novel scheme

success rate than its rivals. In the second two plots, our novel routing scheme outperforms the other two algorithms by creating a lower network delay through the whole simulation time slot.

5.3 Summary and Outlook

The recent advances of cloud computing have substantially changed researcher's understanding of computing hardware/software infrastructure and development methods. This fast transition to cloud computing has enabled a plethora of enterprise services for client use, which has also opened new security challenges. More specifically, serious concerns regarding the confidentiality, integrity, and availability of communication and information systems have arisen as a result of rapid transition to the cloud. For example, relocating data through the communication systems to the clouds has caused various security and privacy issues that did not previously exist in traditional client/server models. This chapter proposed a novel Oblivious Routing-based Security Scheme ($\mathcal{ORSS}$) which effectively addressed the security issues caused by the Distributed Denial of Service (DDoS) security attacks related to the data communication in cloud systems. A detailed theoretical model and the analysis including an experimental work in the form of simulation showed the superior security and performance of the oblivious routing algorithm (utilized by the scheme) compared with conventional routing algorithms widely used today.

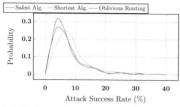

(a) Pdf of attack success rate (percentage) in Oxford network topology.

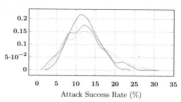

(b) Pdf of attack success rate (percentage) in Missouri network topology.

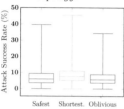

(c) Boxplot of attack success rate in Oxford network topology.

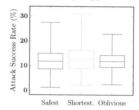

(d) Boxplot of attack success rate in Missouri network topology.

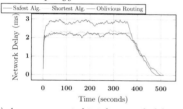

(e) Average expected total network delay over time in Oxford network topology.

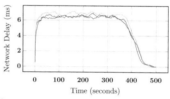

(f) Average expected total network delay over time in Missouri network topology.

Fig. 5.4 Comparison of different routing algorithms based on the average delay and probability distribution of successful attacks. (**a**) Pdf of attack success rate (percentage) in Oxford network topology. (**b**) Pdf of attack success rate (percentage) in Missouri network topology. (**c**) Boxplot of attack success rate in Oxford network topology. (**d**) Boxplot of attack success rate in Missouri network topology. (**e**) Average expected total network delay over time in Oxford network topology. (**f**) Average expected total network delay over time in Missouri network topology

References

1. C. Almond, A practical guide to cloud computing security, Aug 2009
2. B. Joshi, A. Santhana Vijayan, B.K. Joshi, Securing cloud computing environment against DDoS attacks, *2012 International Conference on Computer Communication and Informatics (ICCCI-2012)*, Coimbatore, 10–12 Jan 2012
3. S. Yu, W. Zhou, R. Doss, Information theory based detection against network behavior mimicking DDoS attacks. IEEE Commun. Lett. **12**(4), 318–321 (2008)
4. S. Yu, W. Zhou, Entropy-based collaborative detection of DDoS attacks on community networks, in *Sixth Annual IEEE International Conference on Pervasive Computing and Communications*, Piscataway, NJ, 2008, pp. 566–571
5. J.J.B. Krishnamurthy, M. Rabinovich, Flash crowds and denial of service attacks: characterization and implications for CDNs and web sites, in *Proceedings of International WWW Conferences*, 2002

6. Y. Chen, K. Hwang, Collaborative change detection of DDoS attacks on community and ISP networks, in *Proceedings of IEEE CTS*, 2006
7. R.B.G. Carl, G. Kesidis, S. Rai, Denial of service attack detection techniques. IEEE Internet Comput. **10**(1), 82–89 (2006)
8. B. Wang, Y. Zheng, W. Lou, Y.T. Hou, DDoS attack protection in the era of cloud computing and software-defined networking. Comput. Netw. **81**, 308–319 (2015)
9. A. Mishra, R. Mathur, S. Jain, J. Singh Rathore, Cloud computing security. Int. J. Recent Innov. Trends Comput. Commun. **1**(1), 36–39 (2013)
10. D. Zissis, D. Lekkas, Addressing cloud computing security issues. Futur. Gener. Comput. Syst. **28**(3), 583–592 (2012)
11. S.S. Iyengar, K.G. Boroojeni, *Oblivious Network Routing: Algorithms and Applications* (MIT Press, Cambridge, 2015)
12. A. Gupta, M.T. Hajiaghayi, H. Racke, Oblivious network design, in *SODA '06: Proceedings of the 17th Annual ACM-SIAM Symposium on Discrete Algorithm* (ACM, New York, 2006), pp. 970–979
13. MATLAB version 8.5. Miami, The MathWorks Inc., Florida, 2015
14. S. Knight, H.X. Nguyen, N. Falkner, R. Bowden, M. Roughan, The internet topology zoo. IEEE J. Sel. Areas Commun. **29**(9), 1765–1775 (2011)

Part III
Privacy Preservation

Chapter 6
End-User Data Privacy

Energy systems are undergoing enormous transformations around the world. Though loosely defined, the concept of the smart grid entails power networks transmitting digital information as well as energy. The primary purpose is to allow (near) real-time consumption and generation data to be transmitted between different nodes, but it also allows for possibilities such as remote activation of appliances. In combination with facilitating increased amounts of distributed generation, often from renewable sources with variable output, the goal is to optimize the balance of generation and consumption in order to achieve efficiencies.

Chapter 6 addresses the location privacy concerns that end-users (e.g., smart meters) would have when they use a variety of location-based services on which the smart grid controllers rely for their main functionalities. Section 6.1 introduces the end-users like smart meters and their importance in a given smart grid. Section 6.2 introduces the preliminary concepts of location privacy and the localization attacks. Section 6.3 introduces a couple of location privacy preservation mechanisms as countermeasures to the localization attack. At the end, a summary and outlook of Chap. 6 will be presented.

6.1 Introduction

More than just a grand technological project however, a smart grid has the potential to fundamentally change the social dynamics of the energy system. Two opposing visions of how the smart grid's potential might be realized are established, though they should be considered two poles of a continuum rather than a binary choice. The first, in keeping with the centralized, hierarchical paradigm which has defined the energy systems of the last century, entails centralized generators increasing monitoring and control of end-user consumption, as detailed in UKERC's "Smart Power Sector" scenario of smart grid futures. Henceforth, we refer to this

K.G. Boroojeni et al., *Smart Grids: Security and Privacy Issues*,
DOI 10.1007/978-3-319-45050-6_6

vision as "centralized demand side management" (CDSM), as a specific form of the generic term "demand side management" (DSM). The alternative involves blurring the distinction between generators and end-users, with the latter whether as individuals or communities increasingly independent through microgeneration and self-management, a model Wolsink calls "DisGenMiGrids" (distributed generation microgrids), and similar to UKERC's "Groundswell" scenario. These contrasting visions share the same technologies, but differ radically in the social structures underpinning them.

In extending generator control of consumption, centralized demand side management targets the provision of accurate usage information to consumers, including dynamic pricing tariffs, and the remote control of electricity load and devices. Within these approaches there is considerable latitude in regard to the role envisaged for the user; however, all require integration into daily routines and so some degree of user interaction. A weak version of CDSM might simply entail a smart implementation of dynamic pricing tariffs, in which certain white goods are remotely triggered to run during low demand periods. A strong implementation could include using real-time pricing signals and new technologies to encourage and enable users to time shift energy-intensive behaviors away from periods of peak demand, or towards periods when fluctuating renewable energy generation is high. Such an approach would require energy to take a prominent role in the ordering of household activities.

To date, a considerable body of work has been generated from practice theory-based studies of energy use in domestic contexts. Some of this work has called for a disassembling of the producer-consumer divide which has defined the energy systems of the last century, to be replaced instead by the kind of co-management of resources seen in DisGenMiGrids. Strangers extends this further, calling for co-management of practices, a more ambitious vision which recognizes the co-production of demand that is the relationships in which wants and needs are formed as well as supply. By contrast, much of the work to date on DSM specifically has been narrow in focus and concerned with individual users, disregarding the dynamics of the shared household as a deployment site. Research that has explored this area is often limited to certain aspects of DSM, for example on smart meteres, in home displays (IHDs) and; dynamic pricing; or peer comparison feedback effects.

Researching the societal implications of smart grids faces similar problems to that of other new technologies (e.g., biotechnologies, nanotechnologies) in gaining insight into socio-technical systems that do not yet exist. The uncertainty of future technologies necessitates defining them for research participants. In doing so the context and framing used can have a large influence on responses. Despite this, the necessity of such research stems from the considerable benefit in upstream engagement with new technologies where lay perspectives help to direct research and development efforts, and smart grids are no exception in this regard.

The current research makes use of contravision scenario films within a series of focus groups in order to engage members of the public with the range of potential smart grid technologies available within future energy systems. These enable us to probe people's understandings of, and engagement with, their own energy consumption, and explore interactions with current and future smart grid technologies. Recent work draws attention to the prominent role that the user is expected to play within smart grid systems. Our core interest is what that role might look like, and the consequences of it. Two forms of public are identified energy consumers and energy citizens which in crude terms are distinguished by their orientation: as energy end-users and energy system participants, respectively. It is argued that the energy consumer frame is a consequence of the same paradigm that drives CDSM, and yet it undermines the very thing CDSM hopes to achieve namely a grid in which consumption adjusts to meet generation. We propose that energy citizens, aligning with DisGenMiGrids, hold out much greater potential in this regard. The implications of this for policy makers are discussed.

6.2 Preliminaries to Privacy Preservation Methods

One major concern in the wide deployment of LBSs is how to preserve the location privacy of the sensor nodes while providing them with a service based on their locations. LBS providers can be victims of the privacy attackers to track some sensors, or they themselves may abuse the sensors' location information for malicious purposes [1, 2]. There are two general approaches dealing with the location privacy issue in LBSs. First, we address the couple of approaches and then, we will focus on our contribution.

6.2.1 k-Anonymity Cloaking

In this approach, instead of sending one single node's LBS request to the server, including her exact location, k-anonymity cloaking employs a trusted third party who collects k neighboring nodes' requests and sends them all together to the LBS service provider. This approach doesn't address the case that the node density is high; in this case, the k nodes' locations may be very close to each other, and hence this approach will still reveal the node's location privacy to some extent. This approach was originally proposed by Gruteser and Grunwald [3]. Their work may lead to large service delay if there are not enough nodes requesting LBSs. Later on, Gedik and Liu [4] designed a joint spatial and temporal cloaking algorithm which collects k LBS requests, each from a different node in a specified cloaking area within a specified time period and then sends them to the service provider. A negative point in their work is that if there are only less than k requests within the predefined time period, the nodes' requests will be blocked.

In 2009, Meyerowitz and Choudhury [5] tried to improve the service accuracy by predicting the nodes' paths and LBS queries, and send the results to nodes' before they submit queries. The main drawback of their approach is the network delay occurred because of high communication overhead. For more treatment on k-anonymity cloaking, see [6–11].

6.2.2 *Location Obfuscation*

We can divide the solutions with this approach into two categories: solutions that preserve the location privacy of the nodes by inserting some fake LBS requests; and those which deviate a node's location from the real one in her LBS request to protect her location privacy.

As examples of the solutions in the earlier category, consider the schemes proposed by Kido et al. [12], Lu et al. [13], and Duckham and Kulik [14]. In these schemes, the node generates some fake locations (dummies) using some dummy generation methods and submits the dummies and its own location to the LBS server. The server analyzes every submitted query and replies properly. The major drawback of the solutions in this category is that the server is used inefficiently and may become the system bottle-neck. Additionally, nodes' location privacy is not preserved in advance.

The second category of solutions like the ones proposed by Ardagna et al. [15], Pingley et al. [16], and Damiani et al. [17] hide nodes' real locations, e.g., by submitting shifted locations. Such schemes trade service accuracy for location privacy.

Finally, in 2013, Ming Li et. al. [2] proposed a location privacy preserving scheme called n-CD which doesn't use a third party and provides a trade-off between the privacy level and the system accuracy (concealing cost).

In Sect. 6.3, we hide a node's location in an anonymity zone obtained by the Voronoi diagram of an n-gon in a 2-D plane. The scheme enables the node to specify the minimum level of privacy (λ) that it desires or the maximum error tolerance (ε) that it is willing to accept when informing the services of its location. The level of privacy is defined as the probability that the node's location would not be revealed by the privacy attackers if they know the anonymity zone and the way in which the scheme is constructed. Moreover, the error tolerance is defined as the maximum Euclidean distance between any two points in the anonymity zone in which the node is located. We will theoretically find a trade-off between λ and ε in the following form: the complement of the privacy level is inversely proportional to the cube of the error tolerance; i.e.

$$1 - \lambda = \frac{1}{\Omega(\varepsilon^3)} \tag{6.1}$$

The mentioned relation is obtained for the static node (the case addressed by all the works mentioned in the literature review).

Additionally, in this chapter, we use a random walk in the form of Brownian motion process to model the node's mobility on the plane. To the best of our knowledge, it is the first work that proposes a location privacy-preserving scheme that hides a mobile node's location and movement path in 2-D plane by considering a reasonable stochastic model of his/her movement based on the Brownian Motion process. We theoretically obtain the instantaneous privacy level of the mobile node. The time complexity of our approach is linear ($O(n)$ where n is the number of vertices of the initial polygon).

6.2.3 Preliminary Definitions

Let c denote a mobile node in an Euclidean plane ($\mathbb{R}^2$) that wants to use a location-based service (say A). Assume that c is walking on the plane to do some job; for example, let c be a *sensor* which wants to control some criteria on the 2-D plane. Additionally, assume that there is no obstacle on the plane and node c can go anywhere on the plane (we leave the case with obstacles for future work). We discretize the plane by partitioning it into an infinite countable number of congruent distinct regions. As the result, node c is located in one of the regions at any given moment. Let r_t denote a random process which specifies the region in which c is located at moment $t \geq 0$.

6.2.3.1 Location-Based Service

Assume that node c sends a query message containing its location to location-based service A every one time unit (which can be any value) and gets benefit of its service. Here is the format of the query message that node c sends to A at time $t = n$ (for $n = 0, 1, \ldots$):

$$\mathcal{Q}_c(n) = \langle \text{ID}_c, r_n, \text{DATA} \rangle \tag{6.2}$$

where ID_c specifies the ID of node c, r_n denotes the region where c is located in at time n, and DATA represents the other information that c may need to send to the LBS.

This kind of communication makes the LBS able to keep track of node c during the communication time. This may compromise its privacy (as the LBS or some third party (like data sniffer) may abuse this information). Consequently, we need to define a Location Privacy-Preserving Mechanism (LPPM) to filter the query message before sending it to the LBS. In this paper, we use a location-obfuscation method to filter this information and increase the privacy level of node c.

6.2.3.2 Location Privacy-Preserving Mechanism

In order to preserve the location privacy, node c obfuscates its current location using an LPPM before sending it to the LBS. In fact, instead of sending query $\mathcal{Q}_c(n)$ at time $t = n$, it sends $\mathcal{Q}'_c(n)$ such that

$$\mathcal{Q}'_c(n) = \langle \text{ID}_c, \text{obf}(r_n), \text{DATA} \rangle \tag{6.3}$$

where obf: $\mathcal{R} \mapsto 2^{\mathcal{R}}$ and $\mathcal{R} = \{r_0, r_1, r_2, \ldots\}$ (note that function obf maps every region to a set of regions). In this way, node c sends an obfuscated location each time instead of revealing its exact location to the LBS. This obfuscated location is obtained by the following equation:

$$\text{obfuscated}(r_t) = \bigcup_{\rho \in \text{obf}(r_t)} \rho \qquad (6.4)$$

We call the sequence of regions $\bar{r} = r_0, r_1, r_2, \ldots$ as the *(actual) track* of node c; while the sequence:

$$\text{obfuscated}(r_0), \text{obfuscated}(r_1), \text{obfuscated}(r_2), \ldots$$

is the corresponding *obfuscated* track of c.

6.2.3.3   Adversary

As mentioned before, the LBS itself is considered to be a potential adversary. Additionally, some third party may eavesdrop the query messages originated by node c to use them for some malicious purposes. In this paper, we assume that the adversary knows the LPPM (which is function obf) used by node c to filter the query message. Additionally, at time $t = n$, she is aware of the obfuscated track sent by c in time interval $[0, n]$. Using this information, the adversary wants to reasonably guess the actual track of node c. Let $\hat{\bar{r}} = \hat{r}_0, \hat{r}_1, \ldots$ denote the adversary guess of the actual track such that for every i, $\hat{r}_i \in \mathcal{R}$ specifies the adversary guess of region r_i in which node c is located at time $t = i$.

6.3   Privacy Preservation: Location Obfuscation Methods

Here in this section, we describe a location privacy preserving mechanism which obfuscates the location information of a mobile node c to increase its privacy. In order to specify such a mechanism, we need to define the output value of the aforementioned function *obf* for every given input region $r \in R$.

In this section, we assume that set R partitions the Euclidean plane into an infinite number of unit squares such that:

$$\forall \rho \in R, \exists x, y \in \mathbb{R} : \rho$$
$$= [x, x + 1) \times [y, u + 1)$$

Additionally, we discretize the time space into the set of nonnegative integer (as we have already assumed that the query messages are sent every single time unit).

We describe the mechanism as a function called *region-obfuscator* which gets $\rho \subseteq R$, time $t = n$, and scale factor α as its input and calculates the corresponding obfuscated region and boundary factor (which will be addressed later) as the output. In the case that $n = 0$, the obfuscated region is a square of edge length with parallel edges to *x-y* axes. The square is centered at a point obtained by randomly translating the ρ's centroid with $(x\alpha, y\alpha)$ where x and y are uniformly distributed over interval $(-1/6, 1/6)$ Additionally, the boundary factor is obtained by generating a sample of uniformly distributed random variable in interval $(1/3,1)$.

In the case that $n > 0$, function *region-obfuscator* first needs to check whether the current obfuscated region and boundary factor (which were generated at time $t = n - 1$) are valid for the current region ρ. If yes, it simply returns the previous values; otherwise, it generates a new boundary factor (β) and calls function UPDATE to generate a new region (o).

Here is the validation rule: if region ρ lies inside the square S (which is centered at o'.centroid and has the edge length of $\beta'\alpha$, the current obfuscated region (o') and boundary factor (β') are still valid at time $t = n$.

6.4 Summary and Conclusion

In this chapter, we defined an initial anonymity zone for any static node first. Then, we shrank the zone using some geometric deductions. As mentioned, the shrunk anonymity zone is a convex polygon. We proposed an approach for finding the stochastic distribution of the node over the obtained anonymity zone. Additionally, we described the trade-off existing between the privacy level and the error tolerance of our scheme by obtaining thresholds for both the node's privacy level and the error tolerance. Moreover, we extended our scheme for a mobile node with random walk on the 2D plane. In the mobile version, our scheme guarantees a specified minimum value for the location privacy level while assuring to hide the node's movement path.

References

1. R. Dewri, Location privacy and attacker knowledge: who are we fighting against? in *Security and Privacy in Communication Networks*. Lecture Notes of the Institute for Computer Sciences, Social Informatics and Telecommunications Engineering, vol. 96 (Springer, Berlin, 2012), pp. 96–115
2. M. Li, S. Salinas, A. Thapa, P, Li, *n*-CD: a geometric approach to preserving location privacy in location-based services, in *Proceedings of IEEE INFOCOM*, 2013
3. M. Gruteser, D. Grunwald, Anonymous usage of location-based services through spatial and temporal cloaking, in *ACM Mobisys'03*, May 2003
4. B. Gedik, L. Liu, Protecting location privacy with personalized k-anonymity: architecture and algorithms. IEEE Trans. Mob. Comput. **7**(1), 1–18 (2008)
5. J. Meyerowitz, R.R. Choudhury, Hiding stars with fireworks: location privacy through camouflage, in *Proceedings of ACM MobiCom*, Beijing, Sept 2009

6. M.F. Mokbel, C.Y. Chow, W.G. Aref, The new casper: query processing for location services without compromising privacy, in *Proceedings of VLDB*, 2006
7. P. Kalnis, G. Ghinita, K. Mouratidis, D. Papadias, Preventing location-based identity inference in anonymous spatial queries. IEEE Trans. Knowl. Data Eng. **19**(12), 1719–1733 (2007)
8. B. Gedik, L. Liu, Location privacy in mobile systems: a personalized anonymization model, in *Proceedings of IEEE ICDCS*, Columbus, OH, June 2005
9. C.-Y. Chow, M.F. Mokbel, X. Liu, A peer-to-peer spatial cloaking algorithm for anonymous location-based service, in *Proceedings of ACM GIS*, Arlington, VA, Nov 2006
10. A. Beresford, F. Stajano, Location privacy in pervasive computing. IEEE Pervasive Comput. **2**(1), 46–55 (2003)
11. B. Hoh, M. Gruteser, H. Xiong, A. Alrabady, Preserving privacy in GPS traces via uncertainty-aware path cloaking, in *Proceedings of ACM CCS 2007*, Alexandria, VA, Jan 2007
12. H. Kido, Y. Yanagisawa, T. Satoh, An anonymous communication technique using dummies for location-based services, in *Proceedings of IEEE ICPS*, Santorini, July 2006
13. H. Lu, C.S. Jensen, M.L. Yiu, Pad: privacy-area aware, dummy based location privacy in mobile services, in *Proceedings of ACM MobiDE*, Vancouver, June 2008
14. M. Duckham, L. Kulik, A formal model of obfuscation and negotiation for location privacy, in *Proceedings of International Conference on Pervasive Computing*, Munich, May 2005
15. C.A. Ardagna, M. Cremonini, S.D.C. di Vimercati, P. Samarati, An obfuscation-based approach for protecting location privacy. IEEE Trans. Dependable Secure Comput. **8**(1), 13–27 (2011)
16. A. Pingley, W. Yu, N. Zhang, X. Fu, W. Zhao, Cap: a contextaware privacy protection system for location-based services, in *Proceedings of IEEE ICDCS*, Montreal, June 2009
17. M. Damiani, E. Bertino, C. Silvestri, Probe: an obfuscation system for the protection of sensitive location information in LBS. Technical Report 2001–145, CERIAS, 2008

Chapter 7
Mobile User Data Privacy

Smart grid (SG) concept is introduced to achieve a sustainable, secure, and environmentally-friendly power system by using new elements such as distributed renewable resources, advanced metering infrastructure, and modern transportation in terms of electric vehicle (EV) utilization [1, 2]. In recent years, the U.S. government targets to increase the penetration of modern EVs [3]. From a critical point of view, utilizing large number of EVs connected to the future power grid may threaten the reliability and stability of power grid [4, 5]. The society of automotive engineers (SAE) established some standards about the utilization of EVs including SAE J2847 which institutes requirements and specifications for communication between EVs and power system. This standard specifies interactions between EVs and power system operators [6]. According to [1], from the utilities perspective, it is not elaborately specified whether EV utilization in terms of vehicle to grid (V2G) is cost-effective [7]. In [8], the authors introduced a comparison between direct and deterministic communication structure and proposed an aggregative command transmit architecture considering three influential factors, reliability, availability, and participating EVs in ancillary services.

Chapter 7 addresses the location privacy concerns that mobile components of the modern smart grid (e.g., electric vehicles) would have when they use a variety of location-based services on which the smart grid controllers rely for their main functionalities. Section 7.1 introduces the mobile components in smart grids like electric vehicles and their importance in a given smart grid. Section 7.2 introduces the preliminary concepts of location privacy and the trajectory revealing attacks. Section 7.3 introduces a couple of location privacy preservation mechanisms as countermeasures to such attacks. At the end, a summary and outlook of Chap. 7 will be presented.

© Springer International Publishing Switzerland 2017
K.G. Boroojeni et al., *Smart Grids: Security and Privacy Issues*,
DOI 10.1007/978-3-319-45050-6_7

7.1 Introduction

Based on literature, in the modern transportation context, EVs require communicating information about the state of charge, desired charging rate, and their location data so that charging stations and EV aggregators estimate expected power consumption for next hours and provide an acceptable level of reliable service for EVs. There have been several studies about communication protocols, including ZigBee [9] and Cellular Network which is applicable in long-range wireless systems and EVs' data transfer [10].

There have been several studies about communication protocols, including, but not limited to: ZigBee which is implementable for small mesh networks, has low price, and high redundancy which requires less maintenance [11]; and Cellular Network which is applicable in long-range wireless systems and regarding [12] this type of communication is feasible for EVs' data transfer. Additionally, collaborations between EVs and power system infrastructures, called V2X, are visualized to ameliorate traffic efficiency and driver welfare [9]. In order to implement V2X in a more secure way, there is an exigent need to propose a structure for concealing location and exact driving path of EVs.

Consequently, in the PHEV charging context, the vehicles usually transmit some data, such as identity, state of charge, usage pattern, and location and these data used to be accessed by charging stations [10]. Therefore, one of the main privacy concerns is to protect customers data, especially the vehicles location [13]. The location privacy of EVs includes, but not limited to, drivers home address, working location, and favorite places to travel [14, 15]. Some efforts focused on evaluating the effectiveness of location privacy methods in V2X [16, 17]. In our knowledge, this is the first work which introduces a novel algorithmic structure not only to conceal the location of mobile devices but also to obfuscate their path movement information (Table 7.1).

Knowledge Important data about EV drivers including their location, user ID for each car, current SOC, desired SOC, and car model. Interest of adversary If a third party has access to this data it can be used to make unpredictable peak load in a specific region/feeder in power network Possible adversaries (1) High penetration of EVs means more sensitivity especially in the power system demand estimation considering EVs charging demand. Therefore, accessing to this kind of information can help attackers to increase the demand in a specific region by manipulating transferred data. (2) The exact driving patterns can be extracted based on location information. Therefore it can be used to classify car drivers' behavior by advertising companies without drivers' permission. (3) Location of cars is critical for traffic management and it should not be accessible conveniently by third parties (Fig. 7.1).

Table 7.1 Comparison of different works in the context of location privacy-preserving mechanism

Item	Description
knowledge	Important data about EV drivers including their location, user ID for each car,current SOC, desired SOC, and car model
Interest of adversary	If a third party has access to this data, it can be used to make un-predictable peak load in a specific region/feeder in power network
Possible adversaries	(1) High penetration of EVs means more sensitivity especially in the power system demand estimation considering EV's charging demand . Therefore, accessing to this kind of information can help attackers to increase the demand in a specific region by manipulating the transferred data (2) The exact driving patterns can be extracted based on location information. Therefore it can be used to classify car drivers' behavior by advertising companies without drivers' permission (3) Location of cars is critical for traffic management and it should not be accessible conveniently by third parties

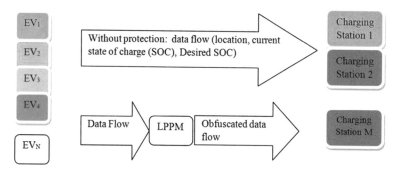

Fig. 7.1 Schematic view of location obfuscation methods in electric vehicle networks

7.2 Preliminaries on Mobile Nodes Trajectory Privacy

In 2011, Shokri et al. [18, 19] formalized localization attacks using Bayesian inference for Hidden Markov Processes. They mathematically modeled a location privacy-preserving mechanism, users mobility pattern, and adversary knowledge-base. They also divide the LBSs into two classes: those who *sporadically* ask their users to expose their location, and those who *continuously* do. Additionally, they quantified the user location privacy as the expected distortion of adversary's guess from the reality of user's location. They used this quantification method to evaluate the effectiveness of location obfuscation and fake location injection mechanisms in the improvement of location privacy level.

Let c denote a mobile node in an Euclidean plane ($\mathbb{R}^2$) that wants to use a location-based service (say A). Assume that c is walking on the plane to do some job; for example, let c be a *sensor* which wants to control some criteria on the 2-D plane. Additionally, assume that there is no obstacle on the plane and node c can go anywhere on the plane (we leave the case with obstacles for future work). We discretize the plane by partitioning it into an infinite countable number of congruent distinct regions. As the result, node c is located in one of the regions at any given moment. Let r_t denote a random process which specifies the region in which c is located at moment $t \geq 0$.

7.2.1 Location-Based Service

Assume that node c sends a query message containing its location to location-based service A every one time unit (which can be any value) and gets benefit of its service. Here is the format of the query message that node c sends to A at time $t = n$ (for $n = 0, 1, \ldots$):

$$Q_c(n) = \langle \mathrm{ID}_c, r_n, \mathrm{DATA} \rangle \tag{7.1}$$

where ID_c specifies the ID of node c, r_n denotes the region where c is located in at time n, and DATA represents the other information that c may need to send to the LBS.

This kind of communication makes the LBS able to keep track of node c during the communication time. This may compromise its privacy (as the LBS or some third party (like data sniffer) may abuse this information). Consequently, we need to define a Location Privacy-Preserving Mechanism (LPPM) to filter the query message before sending it to the LBS. In this paper, we use a location-obfuscation method to filter this information and increase the privacy level of node c.

7.2.2 Location Privacy-Preserving Mechanism

In order to preserve the location privacy, node c obfuscates its current location using an LPPM before sending it to the LBS. In fact, instead of sending query $Q_c(n)$ at time $t = n$, it sends $Q'_c(n)$ such that

$$Q'_c(n) = \langle \mathrm{ID}_c, \mathrm{obf}(r_n), \mathrm{DATA} \rangle \tag{7.2}$$

where $\mathrm{obf}: \mathcal{R} \mapsto 2^{\mathcal{R}}$ and $\mathcal{R} = \{r_0, r_1, r_2, \ldots\}$ (note that function obf maps every region to a set of regions). In this way, node c sends an obfuscated location each time instead of revealing its exact location to the LBS. This obfuscated location is obtained by the following equation:

$$\text{obfuscated}(r_t) = \bigcup_{\rho \in \text{obf}(r_t)} \rho \qquad (7.3)$$

We call the sequence of regions $\bar{r} = r_0, r_1, r_2, \ldots$ as the *(actual) track* of node c; while the sequence:

$$\text{obfuscated}(r_0), \text{obfuscated}(r_1), \text{obfuscated}(r_2), \ldots$$

is the corresponding *obfuscated* track of c.

7.2.3 *Adversary*

As mentioned before, the LBS itself is considered to be a potential adversary. Additionally, some third party may eavesdrop the query messages originated by node c to use them for some malicious purposes. In this paper, we assume that the adversary knows the LPPM (which is function obf) used by node c to filter the query message. Additionally, at time $t = n$, she is aware of the obfuscated track sent by c in time interval $[0, n]$. Using this information, the adversary wants to reasonably guess the actual track of node c. Let $\hat{\bar{r}} = \hat{r}_0, \hat{r}_1, \ldots$ denote the adversary guess of the actual track such that for every i, $\hat{r}_i \in \mathcal{R}$ specifies the adversary guess of region r_i in which node c is located at time $t = i$.

In Sect. 2.3, we obfuscate the movement path (trajectory) and instantaneous location of a randomly walking mobile device in 2-D plane using a novel mechanism. Our approach enables the user to specify the minimum level of privacy (λ) that it desires and the maximum error tolerance (ε) that it is willing to accept when informing the location based services of its location. The level of privacy is defined as the expected error (distortion) of adversary's guess from the user's location. Moreover, the error tolerance is defined as the expected Euclidean distance between the LBSs' guess and user's actual location. We will find a trade-off between λ and ε which will be examined using some simulation results.

7.3 Privacy Preservation Quantification: Probabilistic Model

Here in this section, we describe a novel location privacy-preserving mechanism which obfuscates the location information of mobile node c to increase its privacy. In order to specify such a mechanism, we need to define the output value of the aforementioned function obf for every given input region $r \in \mathcal{R}$.

In this section, we assume that set $\mathcal{R}$ partitions the Euclidean plane into an infinite number of unit squares such that

---

Algorithm 7.1: REGIONOBFUSCATOR

Input: region ρ, time n, & scale factor α
Output: obfuscated region o & boundary factor $\mathcal{B}$
1 **if** $n = 0$ **then**
2 $\quad\quad$ $x \leftarrow \mathcal{U}\mathrm{nif}(-\frac{1}{6}, \frac{1}{6})$; $y \leftarrow \mathcal{U}\mathrm{nif}(-\frac{1}{6}, \frac{1}{6})$;
3 $\quad\quad$ $o \leftarrow$ A square of edge length α and
4 $\quad\quad$ centroid ρ.centroid;
$\quad\quad$ /* edges are parallel to x-y axes $\quad\quad\quad\quad\quad\quad\quad\quad$ */
5 $\quad\quad$ $o \leftarrow$ TRANSLATION$(o, (x\alpha, y\alpha))$;
6 $\quad\quad$ $\mathcal{B} \leftarrow \mathcal{U}\mathrm{nif}(\frac{1}{3}, 1)$;
7 **end**
8 **else**
9 $\quad\quad$ $o' \leftarrow$ the obfuscated region of time $n - 1$;
10 $\quad\quad$ $\mathcal{B}' \leftarrow$ the boundary factor of time $n - 1$;
11 $\quad\quad$ $S \leftarrow$ the square of edge length $\mathcal{B}'\alpha$
12 $\quad\quad$ and centroid $o' \cdot centroid$;
13 $\quad\quad$ **if** $o \subset S$ **then**
14 $\quad\quad\quad\quad$ $o \leftarrow o'$;
15 $\quad\quad\quad\quad$ $\mathcal{B} \leftarrow \mathcal{B}'$;
16 $\quad\quad$ **end**
17 $\quad\quad$ **else**
18 $\quad\quad\quad\quad$ $o \leftarrow$ UPDATE(ρ, o');
19 $\quad\quad\quad\quad$ $\mathcal{B} \leftarrow \mathcal{U}\mathrm{nif}(\frac{1}{3}, 1)$
20 $\quad\quad$ **end**
21 **end**
22 **return** $(o, \mathcal{B})$;

$$\forall \rho \in \mathcal{R}, \exists x, y \in \mathbb{R} : \rho = [x, x + 1) \times [y, y + 1) \tag{7.4}$$

Additionally, we discretize the time space into the set of non-negative integer (as we have already assumed that the query messages are sent every single time unit).

Algorithm 7.1 specifies how the mechanism works.

As you see, function REGIONOBFUSCATOR gets region $\rho \in \mathcal{R}$, time $t = n$, and scale factor α as its input and calculates the corresponding obfuscated region and *boundary factor* (which will be addressed later) as the output.

In the case that $n = 0$, the obfuscated region is a square of edge length α with parallel edges to x-y axes. The square is centered at a point obtained by randomly translating the ρ's centroid:

$$o \cdot \text{centroid} = \rho \cdot \text{centroid} + (x\alpha, y\alpha) \tag{7.5}$$

where x and y are uniformly distributed over interval $(-1/6, 1/6)$. Additionally, the boundary factor is obtained by generating a sample of random variable $\mathcal{U}\mathrm{nif}(1/3, 1)$.

As you see in Algorithm 7.2, in the case that $n > 0$, function REGIONOBFUSCA-TOR first needs to check whether the current obfuscated region and boundary factor (which were generated at time $t = n - 1$) are *valid* for the current region ρ. If yes, it simply returns the previous values; otherwise, it generates a new boundary factor ($\mathcal{B}$) and calls function UPDATE to generate a new region (o).

Here is the validation rule: if region ρ lies inside the square $\mathcal{S}$ (which is centered at o'.centroid and has the edge length of $\mathcal{B}'\alpha$), the current obfuscated region (o') and boundary factor ($\mathcal{B}'$) are still valid at time $t = n$.

Now, we consider the case that the validation rule doesn't hold. In other words, node c is no longer inside square $\mathcal{S}$. Let l denote the line segment connecting points r_{n-1}·centroid and r_n·centroid (note that $\rho = r_n$). Assuming that $\mathcal{B}' > 2/3$, there are two possible cases:

Case 1: If l intersects b_i, region o will be the area inside a square of edge length α and centroid M'_i (for every $i = 1, 2, 3, 4$).

Case 2: If l intersects b'_i, region o will be the area inside a square of edge length α and centroid N_i (for every $i = 1, 2, 3, 4$). Assuming that random variable Z is uniformly distributed over interval $(-1/6, 1/6)$, point $N_i = (X_i, Y_i)$ is obtained by the following equations:

$$X_i = \begin{cases} \rho \cdot \text{centroid} \cdot x + \alpha Z & i = 1 \\ o' \cdot \text{centroid} \cdot x + \frac{\alpha}{3} & i = 2 \\ \rho \cdot \text{centroid} \cdot x + \alpha Z & i = 3 \\ o' \cdot \text{centroid} \cdot x - \frac{\alpha}{3} & i = 4 \end{cases} \tag{7.6}$$

$$Y_i = \begin{cases} o' \cdot \text{centroid} \cdot y + \frac{\alpha}{3} & i = 1 \\ \rho \cdot \text{centroid} \cdot y + \alpha Z & i = 2 \\ o' \cdot \text{centroid} \cdot y - \frac{\alpha}{3} & i = 3 \\ \rho \cdot \text{centroid} \cdot y + \alpha Z & i = 4 \end{cases} \tag{7.7}$$

In addition, if $\mathcal{B}' \leq 2/3$, there exists only one possible case (Case 2), as $b_i = \emptyset$ for every $i = 1, 2, 3, 4$.

We implement our proposed LPPM using MATLAB R2013a to evaluate its performance in practice. Figure 7.2 illustrates how the LPPM hides the instantaneous location and path movement of a randomly walking node. In order to find the value of instantaneous privacy level, we use the recent equation. In this specific case, $T_1 = 162$, $T_2 = 179$. Figures 7.3 and 7.4, respectively, show the plot of instantaneous privacy level in time intervals $[0, 161]$ and $[162, 178]$. As you see, the value of privacy level becomes very close to one at the first few time units. Additionally, after every update, the instantaneous privacy level drops to $1 - 9/\alpha^2 = 0.9955$; i.e. the minimum privacy level is $1 - 9/\alpha^2 = 1 - \Theta(\varepsilon^{-2})$ (trade-off between λ and ε). Figure 7.5 depicts the plot of function $\log(1 - \lambda)$ over time interval $[162, 178]$. As you see, the privacy level gets close to one with *exponential* rate (as $\log(1 - \lambda)$ linearly grows).

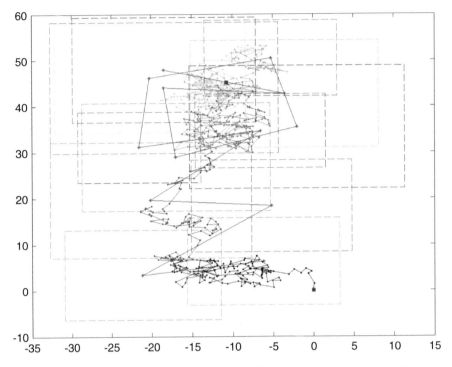

Fig. 7.2 Track of node c walking randomly on the Euclidean plane (x-y Cartesian coordinates) in time interval $[0, 700]$. In this case, $\alpha = 45$, $\sigma^2 = 1$, and $(x_0, y_0) = (0, 0)$. As you see, the track of node c has been shown with *black/gray color* (the *darker*, the earlier). Additionally, the start and end point of the track has been specified with *asterisk and plus sign*, respectively. *Dashed squares* show the boundary of obfuscated region at different times (edge length of each *square* is $\mathcal{B}\alpha$ where $\mathcal{B}$ is the boundary factor). The *red track* specifies the obfuscated track which connects the centroids of subsequent *squares*

7.4 A Vernoi-Based Location Obfuscation Method

This section presents an extension of the Vornoi-based scheme, described in Chap. 6, for obfuscating the location and trajectory of a mobile node. We restrict our consideration to the case that the node has subsequent random close by destinations. More precisely, the node has a sequence of independently chosen random destinations in the form $\mathcal{D}_0 = \text{loc}_c(0), \mathcal{D}_1, \mathcal{D}_2, \mathcal{D}_3, \ldots$ such that for every $i \geq 0$, the node moves from point $\mathcal{D}_i \in \mathbb{R}^2$ to $\mathcal{D}_{i+1} \in \mathbb{R}^2$ in time interval $[t_i, t_{i+1})$ such that $t_0 = 0$,

$$0 < t_{i+1} - t_i \leq \epsilon \qquad \forall i = 0, 1, \ldots$$

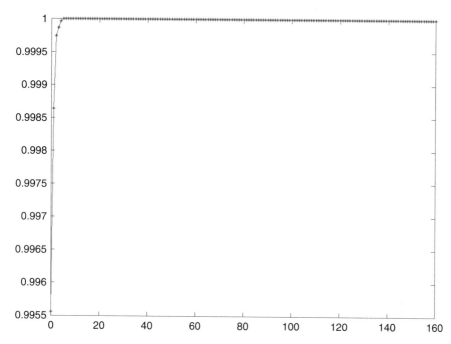

Fig. 7.3 Plot of instantaneous privacy-level over time interval $[0, 161]$ (before the first update). y axis specifies $\lambda(t, \bar{\rho})$ while $x = t$

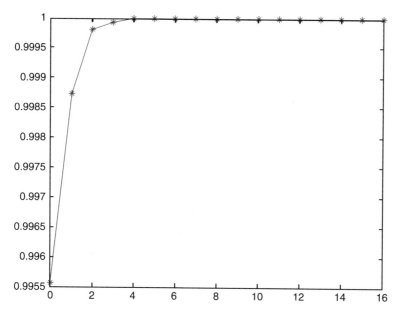

Fig. 7.4 Plot of instantaneous privacy-level over time interval $[162, 178]$ (after the first update and before the second one). y axis specifies $\lambda(t, \bar{\rho})$ while $x = t - 162$

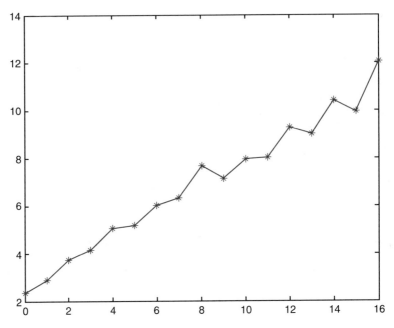

Fig. 7.5 Plot of function $\log_{10}(1 - \lambda)$ over time interval $[162, 178]$. y axis specifies $\log_{10}(1 - \lambda(t, \bar{\rho}))$ while $x = t - 162$

for some small real number $\epsilon > 0$, and assuming that the maximum speed of the node is represented by M_s, this is the case that

$$\epsilon M_s \ll \mu$$

or equivalently, the distance between two consequent destinations $||\mathcal{D}_{i+1} - \mathcal{D}_i||$ is negligible (compared to the scale factor).

7.4.1 A Stochastic Model of the Node Movement

We define random processes x_t and y_t in the following form:

$$\begin{cases} x_t = x_c(t) - x_c(0) \\ y_t = y_c(t) - y_c(0) \end{cases} \quad \forall t \geq 0 \qquad (7.8)$$

where $\big(x_c(t), y_c(t)\big)$ denotes the Cartesian coordinates of the node's location in time t ($\text{loc}_c(t)$).

Considering the aforementioned assumptions regarding the node movement on the plane, we estimate x_t and y_t using random processes $\hat{x}_t$ and $\hat{y}_t$ which have the following properties:

1. Maps $g : t \mapsto \hat{x}_t(\omega)$ and $g' : t \mapsto \hat{y}_t(\omega)$ are continuous for every ω and $t > 0$.
2. For every $k \in \mathbb{N}$, assuming that $0 \leq t_1 \leq t_2 \leq \ldots \leq t_k$, the random variables belonging to the following set are mutually independent:

$$\{(\hat{x}_{t_{i+1}} - \hat{x}_{t_i}) | i = 1 \ldots k - 1\}$$

The same proposition is true for the following set:

$$\{(\hat{y}_{t_{i+1}} - \hat{y}_{t_i}) | i = 1 \ldots k - 1\}$$

3. Every increment of processes $\hat{x}_t$ and $\hat{y}_t$ is stationary; i.e., the probability distributions of $\hat{x}_t - \hat{x}_s$ and $\hat{y}_t - \hat{y}_s$ only depend on $t - s$ for every $t, s > 0$.

Note that since $loc_c(t) = (x_t + x_c(0), y_t + y_c(0))$, the first mentioned property is also true for x_t and y_t:

$$\begin{cases} x_t = \lim_{w \to t^+} x_w = \lim_{w \to t^-} x_w \\ y_t = \lim_{w \to t^+} y_w = \lim_{w \to t^-} y_w \end{cases}$$

However, the other two properties are reasonably estimated regarding processes x_t and y_t.

Random processes $\hat{x}$ and $\hat{y}$ are known as Brownian Motion processes and this is the case that

$$\begin{cases} \hat{x}_t \sim \mathcal{N}(0, \sigma^2 t) \\ \hat{y}_t \sim \mathcal{N}(0, \sigma^2 t) \end{cases} \tag{7.9}$$

Regarding Eq. (7.8), we find an estimation stochastic process for $x_c(t)$ and $y_c(t)$:

$$\begin{cases} x_c(t) \sim \mathcal{N}(x_c(0), \sigma^2 t) \\ y_c(t) \sim \mathcal{N}(y_c(0), \sigma^2 t) \end{cases} \tag{7.10}$$

7.4.2   Proposed Scheme for a Mobile Node

Now, we extend our scheme to the case that node c is assumed to be moving in the way mentioned previously.

Algorithm 7.2 proposes an appropriate procedure which generates an anonymity zone for mobile node c at time $t = 0$ and keeps it updated as the node is moving for every $t \geq 0$. In this procedure, we assume that function AZGEN-ERATOR(O, n, μ, γ_n) works in the following way: consider Algorithm 7.2 which generates the initial anonymity zone for a static node. If we change the second line of this algorithm to the form of Expression (7.11), we obtain another function called MOBILEZONEGENERATOR(O, n, μ, γ_n) where γ_n is a positive real number less than $2\sin^2(\frac{\pi}{n})$. Function AZGENERATOR returns the zone generated by applying the greedy algorithm mentioned in Sect. 7.3 on the output zone of function MOBILEZONEGENERATOR.

$$d \leftarrow \text{Unif}\left(\mu\left(\cos\left(\frac{2\pi}{n}\right) + \gamma_n\right), \mu\right) \qquad (7.11)$$

Algorithm 7.2: MOBILEZONEUPDATER

Input: privacy level λ & odd integer $n \geq 3$ & scale factor $\mu > 0$ &
$\quad\quad \gamma_n < 2\sin^2(\frac{\pi}{n})$

1 $S_c \leftarrow$ AZGENERATOR$(\text{loc}_c(0), n, \mu, \gamma_n)$;
2 **while true do**
3 $\quad$ $t \leftarrow$ Now();
$\quad\quad$ // Function Now() returns the current time $t \geq 0$.
4 $\quad$ **if** $\text{loc}_c(t) \in S_c$ **then**
5 $\quad\quad$ | **continue**;
6 $\quad$ **end**
7 $\quad$ $S_c \leftarrow$ AZGENERATOR$(\text{loc}_c(t), n, \mu, \gamma_n)$;
8 **end**

7.4.3   Computing the Instantaneous Privacy Level

Now, we find a lower-bound for the node's privacy level at any given time $t > 0$. Without loss of generality, we assume that the anonymity zone S_c has been generated at time $t = 0$ and kept unchanged in time interval $[0, T]$.

$$\mathbf{Pr}\left[G \in B\left(\text{loc}_c(t), r\right) \middle| \text{loc}_c(t) \in S_c\right] = \iint\limits_{\substack{(x_0, y_0) \\ \in S_c}} \mathbf{Pr}\left[G \in B\left(\text{loc}_c(t), r\right) \middle| \text{loc}_c(0) = (x_0, y_0)\right.$$

$$\wedge \text{loc}_c(t) \in S_c \Big] \times \mathbf{Pr}\left[\text{loc}_c(0) = (x_0, y_0)\right] dx_0 dy_0$$

$$(7.12)$$

Additionally, considering the estimated stochastic model of $loc_c(t) = (x_c(t), y_c(t))$ in Relation (7.10), we obtain the following relations:

$$\mathbf{Pr}\left[G \in B(loc_c(t), r) \,\middle|\, \begin{matrix} loc_c(0) = (x_0, y_0) \\ \wedge loc_c(t) \in S_c \end{matrix}\right]$$

$$= \mathbf{Pr}\left[loc_c(t) \in B(G, r) \,\middle|\, \begin{matrix} loc_c(0) = (x_0, y_0) \\ \wedge loc_c(t) \in S_c \end{matrix}\right]$$

$$= \iint\limits_{\substack{(x,y) \\ \in B(G,r)}} \mathbf{Pr}\left[(x_c(t), y_c(t)) = (x, y) \,\middle|\, (x_c(t), y_c(t)) \in S_c\right]$$

$$\leq \frac{\iint\limits_{\substack{(x,y) \\ \in B(G,r)}} \mathbf{Pr}\left[(x_c(t), y_c(t)) = (x, y)\right]}{\mathbf{Pr}\left[(x_c(t), y_c(t)) \in S_c\right]}$$

Since for every $t \leq T$, $x_c(t)$ and $y_c(t)$ are normal distributed random variables of mean x_0 and y_0, respectively, this is the case that $(O = loc_c(0) = (x_0, y_0))$:

$$\iint\limits_{\substack{(x,y) \\ \in B(G,r)}} \mathbf{Pr}\left[(x_c(t), y_c(t)) = (x, y)\right] \leq \iint\limits_{\substack{(x,y) \\ \in B(O,r)}} \mathbf{Pr}\left[(x_c(t), y_c(t)) = (x, y)\right]$$

which implies that

$$\mathbf{Pr}\left[G \in B(loc_c(t), r) \,\middle|\, \begin{matrix} loc_c(0) = (x_0, y_0) \\ \wedge loc_c(t) \in S_c \end{matrix}\right] \leq \frac{\iint\limits_{\substack{(x,y) \\ \in B(O,r)}} \mathbf{Pr}\left[(x_c(t), y_c(t)) = (x, y)\right]}{\mathbf{Pr}\left[(x_c(t), y_c(t)) \in S_c\right]}$$

$$\leq \frac{\int_{r'=0}^{r} \int_{\theta=0}^{2\pi} \frac{r'}{2\pi\sigma^2} \cdot e^{-\frac{r'^2}{2\sigma^2 t}} dr' d\theta}{\mathbf{Pr}\left[(x_c(t), y_c(t)) \in S_c\right]}$$

$$\tag{7.13}$$

Here, we make a claim which will be proved later:

$$B(O, r_{min}) \subseteq S_c \tag{7.14}$$

where

$$r_{min} = \frac{\mu \gamma_n}{2 \sin\left(\frac{2\pi}{n}\right)} \tag{7.15}$$

Regarding Relation (7.14), we obtain the following inequality:

$$
\mathbf{Pr}\left[G \in B\big(\mathrm{loc}_c(t), r\big) \,\middle|\, \begin{array}{l}\mathrm{loc}_c(0) = (x_0, y_0) \\ \wedge \mathrm{loc}_c(t) \in S_c\end{array}\right] \leq \frac{\int_{r'=0}^{r}\int_{\theta=0}^{2\pi} \frac{r'}{\sqrt{2\pi t\sigma^2}} \cdot e^{-\frac{r'^2}{2t\sigma^2}}\, dr'd\theta}{\mathbf{Pr}\left[\big(x_c(t), y_c(t)\big) \in B\big(O, r_{\min}\big)\right]}
$$

$$
\leq \frac{\int_{r'=0}^{r}\int_{\theta=0}^{2\pi} \frac{r'}{\sqrt{2\pi t\sigma^2}} \cdot e^{-\frac{r'^2}{2t\sigma^2}}\, dr'd\theta}{\int_{r'=0}^{r_{\min}}\int_{\theta=0}^{2\pi} \frac{r'}{\sqrt{2\pi t\sigma^2}} \cdot e^{-\frac{r'^2}{2t\sigma^2}}\, dr'd\theta}
$$

$$
\leq \frac{1 - e^{-\frac{r^2}{2t\sigma^2}}}{1 - e^{-\frac{r_{\min}^2}{2t\sigma^2}}}
\tag{7.16}
$$

Using Eq. (7.12) and Inequality (7.16), we conclude that

$$
\mathbf{Pr}\left[G \in B\big(\mathrm{loc}_c(t), r\big) \,\middle|\, \mathrm{loc}_c(t) \in S_c\right]
$$

$$
\leq \frac{1 - e^{-\frac{r^2}{2t\sigma^2}}}{1 - e^{-\frac{r_{\min}^2}{2t\sigma^2}}} \times \iint\limits_{\substack{(x_0, y_0) \\ \in S_c}} \mathbf{Pr}\left[\mathrm{loc}_c(0) = (x_0, y_0)\right] dx_0 dy_0
\tag{7.17}
$$

We replace $\mathbf{Pr}\left[\mathrm{loc}_c(0) = (x_0, y_0)\right]$ by its upper-bound in Inequality (7.17):

$$
\mathbf{Pr}\left[G \in B\big(\mathrm{loc}_c(t), r\big) \,\middle|\, \mathrm{loc}_c(t) \in S_c\right]
$$

$$
\leq \frac{1 - e^{-\frac{r^2}{2t\sigma^2}}}{1 - e^{-\frac{r_{\min}^2}{2t\sigma^2}}} \times \iint\limits_{\substack{(x_0, y_0) \\ \in S_c}} \mathbf{Pr}\left[\bigwedge_{i=0}^{m-1} X_i' = \mathrm{dist}((x_0, y_0), l_i')\right] dx_0 dy_0
\tag{7.18}
$$

where line l_i' and function dist are defined as the same as what mentioned previously. In addition, similar to the proof of recent claim, we can get the following inequality:

$$
\mathbf{Pr}\left[\bigwedge_{i=0}^{m-1} X_i' = \mathrm{dist}((x_0, y_0), l_i')\right] \leq \zeta_n\left(\frac{r}{\mu}\right)^3
\tag{7.19}
$$

for some real positive sequence ζ_n. Inequalities (7.18) and (7.19) imply that

$$
\mathbf{Pr}\left[G \in B\big(\mathrm{loc}_c(t), r\big) \,\middle|\, \mathrm{loc}_c(t) \in S_c\right] \leq \frac{1 - e^{-\frac{r^2}{2t\sigma^2}}}{1 - e^{-\frac{r_{\min}^2}{2t\sigma^2}}} \times \zeta_n\left(\frac{r}{\mu}\right)^3 |S_c|
$$

It is easy to see that the recent claim is also true for the mobile case; i.e. area S_c belongs to ball $B\left(O, \mu \sin(\frac{\pi}{n})\right)$. As the result, this is the case that

$$|S_c| \le \left| B\left(O, \mu \sin\left(\frac{\pi}{n}\right)\right)\right|$$

$$\le \pi\mu^2 \sin^2\left(\frac{\pi}{n}\right)$$

Henceforth, we obtain a lower-bound for the instantaneous privacy level of mobile node c:

$$\lambda(t) \le \pi\mu^2 \sin^2\left(\frac{\pi}{n}\right) \zeta_n \frac{1 - e^{-\frac{r^2}{2t\sigma^2}}}{1 - e^{-\frac{r_{min}^2}{2t\sigma^2}}} \times \left(\frac{r}{\mu}\right)^3$$

or,

$$\lambda(t) \le \zeta_n' \frac{1 - e^{-\frac{r^2}{2t\sigma^2}}}{1 - e^{-\frac{r_{min}^2}{2t\sigma^2}}} \times \left(\frac{r^3}{\mu}\right) \tag{7.20}$$

for some positive real sequence ζ_n'.

To complete our analysis, we need to show Relation (7.14). Remember the notation X_i' which specifies the Euclidean distance between the static node's location O and the ith edge of the polygon S_c for every $i = 0 \ldots m-1$. The change we made in this algorithm will increase the minimum possible value of random variable X_i' from zero to r_{min}:

$$r_{min} = \frac{\mu \gamma_n}{2 \sin(\frac{2\pi}{n})}$$

Henceforth, we conclude Relation (7.14).

7.4.4 Concealing the Movement Path

In order to preserve the location privacy of a mobile node, not only we need to hide its instantaneous location, but we have to conceal its movement path in some extent. In our stochastic scheme, we quantify the privacy level of the node's path by calculating a probabilistic low-threshold for random variable T that is the length of the time interval in which function MOBILEZONEUPDATER (Algorithm 7.2) keeps the anonymity zone unchanged.

As mentioned before, $B(O, r_{min}) \subseteq S_c$. This implies that

$$\sup_{t \le t'} \{||loc_c(t) - loc_c(0)||\} \le r_{min} \to T \ge t'$$

Subsequently, we obtain the following proposition:

$$\sup_{t \leq t'} \{x_c(t) - x_c(0)\} \leq \frac{r_{\min}}{\sqrt{2}} \wedge \sup_{t \leq t'} \{y_c(t) - y_c(0)\} \leq \frac{r_{\min}}{\sqrt{2}}$$

$$\rightarrow T \geq t' \tag{7.21}$$

Now, we defined processes M_t and M_t' in the following form:

$$\begin{cases} M_{t'} = \sup_{t \leq t'} \{x_c(t) - x_c(0)\} \\ \\ M_{t'}' = \sup_{t \leq t'} \{y_c(t) - y_c(0)\} \end{cases} \tag{7.22}$$

Concerning Proposition (7.21), we obtain the following inequality:

$$\mathbf{Pr}[T \geq t'] \geq \mathbf{Pr}\left[M_{t'} \leq \frac{r_{\min}}{\sqrt{2}}\right] \times \mathbf{Pr}\left[M_{t'}' \leq \frac{r_{\min}}{\sqrt{2}}\right] \tag{7.23}$$

Processes M_t and M_t', respectively, represent the running maximum[1] of processes $(x_c(t) - x_c(0))$ and $(y_c(t) - y_c(0))$ which has been previously estimated by two Brownian motion processes of variance σ^2. As the result, this is the case that

$$\begin{cases} \mathbf{Pr}[M_t \leq m] = \mathrm{erf}\left(\frac{m}{\sqrt{2t\sigma^2}}\right) \\ \\ \mathbf{Pr}[M_t' \leq m] = \mathrm{erf}\left(\frac{m}{\sqrt{2t\sigma^2}}\right) \end{cases} \tag{7.24}$$

Consequently, we obtain a probabilistic low-threshold for random variable T:

$$\mathbf{Pr}[T \geq t'] \geq \mathrm{erf}^2\left(\frac{r_{\min}}{2\sqrt{t'\sigma^2}}\right) \tag{7.25}$$

7.5 Summary and Conclusion

In this chapter, we addressed the location privacy of mobile devices connected to smart grids, especially a randomly walking node on the Euclidean plane which continuously exposes its location to an LBS (or possibly an adversary). Then,

---

[1]Running maximum M_t of the Brownian Motion process B_t is a random process which has the following cumulative density function at the arbitrary time $t > 0$ (σ^2 represents the variance of process B_t): $F_{M_t}(m) = \mathrm{erf}\left(\frac{m}{\sqrt{2t\sigma^2}}\right)$ for every $m \geq 0$.

we quantified the privacy-level of the (sensor) node over time by computing the expected distortion of adversary's guess from the reality of node's trajectory. Additionally, the trade-off between the privacy level and maximum error tolerance of the mobile node was examined closely. As a result, the minimum privacy level occurs at the moments that the obfuscated location is being updated and we obtained that $1 - \lambda_{min} = \Theta(\varepsilon_{max}^{-2})$. Finally, we used some simulations to support our theoretical results and prove the efficacy of our method in practice.

References

1. R. Dewri, Location privacy and attacker knowledge: who are we fighting against? in *Security and Privacy in Communication Networks*. Lecture Notes of the Institute for Computer Sciences, Social Informatics and Telecommunications Engineering, vol. 96 (Springer, Berlin, 2012), pp. 96–115
2. M. Li, S. Salinas, A. Thapa, P. Li, *n*-CD: a geometric approach to preserving location privacy in location-based services, in *Proceedings of IEEE INFOCOM*, 2013
3. M. Gruteser, D. Grunwald, Anonymous usage of location-based services through spatial and temporal cloaking, in *ACM Mobisys'03*, May 2003
4. B. Gedik, L. Liu, Protecting location privacy with personalized k-anonymity: architecture and algorithms. IEEE Trans. Mob. Comput. **7**(1), 1–18 (2008)
5. J. Meyerowitz, R.R. Choudhury, Hiding stars with fireworks: location privacy through camouflage, in *Proceedings of ACM MobiCom*, Beijing, Sept 2009
6. M.F. Mokbel, C.Y. Chow, W.G. Aref, The new casper: query processing for location services without compromising privacy, in *Proceedings of VLDB*, 2006
7. C.-Y. Chow, M.F. Mokbel, X. Liu, A peer-to-peer spatial cloaking algorithm for anonymous location-based service, in *Proceedings of ACM GIS*, Arlington, VA, Nov 2006
8. A. Beresford, F. Stajano, Location privacy in pervasive computing. IEEE Pervasive Comput. **2**(1), 46–55 (2003)
9. B. Hoh, M. Gruteser, H. Xiong, A. Alrabady, Preserving privacy in GPS traces via uncertainty-aware path cloaking, in *Proceedings of ACM CCS 2007*, Alexandria, VA, Jan 2007
10. H. Kido, Y. Yanagisawa, T. Satoh, An anonymous communication technique using dummies for location-based services, in *Proceedings of IEEE ICPS*, Santorini, July 2006
11. P. Kalnis, G. Ghinita, K. Mouratidis, D. Papadias, Preventing location-based identity inference in anonymous spatial queries. IEEE Trans. Knowl. Data Eng. **19**(12), 1719–1733 (2007)
12. B. Gedik, L. Liu, Location privacy in mobile systems: a personalized anonymization model, in *Proceedings of IEEE ICDCS*, Columbus, OH, June 2005
13. H. Lu, C.S. Jensen, M.L. Yiu, Pad: privacy-area aware, dummy based location privacy in mobile services, in *Proceedings of ACM MobiDE*, Vancouver, June 2008
14. M. Duckham, L. Kulik, A formal model of obfuscation and negotiation for location privacy, in *Proceedings of International Conference on Pervasive Computing*, Munich, May 2005
15. C.A. Ardagna, M. Cremonini, S.D.C. di Vimercati, P. Samarati, An obfuscation-based approach for protecting location privacy. IEEE Trans. Dependable Secure Comput. **8**(1), 13–27 (2011)
16. A. Pingley, W. Yu, N. Zhang, X. Fu, W. Zhao, Cap: a contextaware privacy protection system for location-based services, in *Proceedings of IEEE ICDCS*, Montreal, June 2009
17. M. Damiani, E. Bertino, C. Silvestri, Probe: an obfuscation system for the protection of sensitive location information in LBS. Technical Report 2001–145, CERIAS, 2008

18. R. Shokri, G. Theodorakopoulos, C. Troncoso, J.-P. Hubaux, J.-Y. Le Boudec, Protecting location privacy: optimal strategy against localization attacks, in *CCS '12 Proceedings of the 2012 ACM conference on Computer and Communications Security*, New York, NY, 2012, pp. 617–627
19. R. Shokri, G. Theodorakopoulos, J.-Y. Le Boudec, J.-P. Hubaux, Quantifying location privacy, in *2011 IEEE Symposium on Security and Privacy (SP)*, Berkeley, CA, May 2011, pp. 247–262

Index

A

Active power, 5, 6, 32–37, 39
Adequacy, 19, 22–25, 39, 53
Advanced metering infrastructure (AMI), 35, 93
Adversary, 8, 10, 74, 90, 94, 95, 97, 108, 109
Adversary model, 95
Aikaike/Bayesian information criteria (AIC/BIC), 55, 64, 66, 68
Anonymity zone, 88, 91, 104, 107
ARMA. *See* Auto-regressive moving average (ARMA)
Auto-correlation function (ACF), 54, 55, 57–64
Auto-regressive moving average (ARMA), 61, 63–64, 66

B

Bad data detection, 53–68
Bayesian-based estimator, 39–42
Blocking data flow, 74
Box–Jenkins, 54, 57, 65

C

Cloud computing, 6, 31, 71, 72, 77, 80
Cloud network, 71–81
Communication network, 1, 36, 74
Communication protocol security, 4
Complex dynamical network, 4, 31
Congestion prevention, 19, 24–25
Contingency analysis, 5, 32, 33, 35
Covariance matrix, 38, 39, 41, 42

D

Data privacy, 10, 11, 85–91, 93–109
Data security, 71–81
Data traffic, 6, 7, 72
DC power flow, 5–6, 31–48
Demand, 1–4, 8, 19–25, 35, 56, 86, 94, 95
Demand response, 2, 3, 31, 35
Detection mechanism, 7
Distributed denial of service (DDoS), 6, 7, 71–74, 79, 80
Distributed random variable, 91, 105
Driver, 94, 95

E

Economic dispatch, 2, 20–21, 24–27
Electric vehicles, 2–4, 19, 21, 31, 93, 95
End-user, 11, 35, 85–91
Energy consumption, 3, 31, 87
Energy systems, 85–87
Error detection, 31–48
Euclidean distance, 88, 97, 107
Experimental work, 80

F

Flexibility, 4, 77
Forecaster, 2, 59, 63–65, 68

G

Geometric deduction, 91

H

Hierarchical decomposition sequence (HDS), 76, 77

© Springer International Publishing Switzerland 2017
K.G. Boroojeni et al., *Smart Grids: Security and Privacy Issues*,
DOI 10.1007/978-3-319-45050-6

Printed in the United States
By Bookmasters